Planes and Pilots

FRENCH AIRCRAFT

1939 - 1942

Fighters, Bombers, Reconnaissance and Observation Types

VOLUME 2 : From Dewoitine to Potez

André JOUINEAU
Dominique BREFFORT

Translated from the French by Alan McKay

HISTOIRE & COLLECTIONS

FOREWORD

This is the eighth volume in our series Planes and Pilots, and the second volume of 'French Aircraft 1939-1945'. Presented in alphabetical order, it evokes some of the most famous machines of the period, famous either because they were produced in large numbers and so there were many of them (like the MS 406, or the Potez 63 and its derivatives - more than one thousand examples of each of these machines coming off the production lines), or because they were the best that the Armée de l'Air had at the time (like the LeO 451 or the almost legendary Dewoitine D.520) even though they often arrived too late to have any great effect on the outcome of the fighting. The following pages again show the great variety of the materiel used; it often had good performances, but it was all too often out of date - this was the major failing which the crews had increasingly to make up for every day with their experience and their courage, seeing as the military situation was getting worse and worse and material conditions were deteriorating daily.

© SHHA

NB. As with the first volume, for simplicity's sake we have decided to arrange the chapters in alphabetical order, the planes being presented under the names they were most commonly known by, even though these may not correspond to those given in the official texts. So for instance, the MS 406 will be found under the title Morane-Saulnier rather than the more official SNCAO.

Dewoitine 500, 501 & 510

When Emile Dewoitine proposed his D.500 fighter at the end of the thirties for the C1 programme (single-seat fighter) he already had a solid reputation as a builder of monoplane fighters behind him. The first to bear his name was the Dewoitine D.1, built in 1931 of which 200 were produced and used by several air forces. Through the intermediary of a new company, the SAF (Société Aéronautique Française- Avions Dewoitine) established in 1928, he brought out his next prototype, the D.38, the following year. Although this machine did not run to a production series it did serve as a test bed for the D. 50 and 50bis models at the beginning of 1931. It was the latter aircraft, redesigned and renamed **D.500** which was put forward for the 1930 C1 competition which specified a monoplane fighter weighing between 3 740 lb and 3 960 lb, armed with at least two MAC 1934 machine guns (in fact Vickers machine guns made under licence in France) able to climb to 33 000 feet with a top speed of 205 mph at 11 550 feet.

The D.500's first flight took place in June 1932 although no official order had been placed - this only happened in November of the following year, after the plane had been declared the winner of the competition. Technically speaking, the D.500 benefited from several innovations and its design was even daring from certain points of view; it had cantilever wings and it was made entirely of metal (including the moving surfaces), at a time when most builders were satisfied with a mixed construction of wood and metal and high, so-called 'parasol' wings, or even two wings (biplane or sesquiplane). This technical audacity together with a Hispano-Suiza 12 Xbrs engine boosted by a supercharger and rated at 690 bhp at 15 840 feet enabled it to reach 225 mph which meant that it flew 62 mph faster than the Nieuport-Delage 62/622! Dewoitine's production capacity being limited, it was the Lioré et Olivier factory which was chosen to produce most of the machines, but because deliveries of the engines were unpredictable, it was only from April 1935 that the first examples could be supplied.

After 100 production machines had been made, with three sold to Venezuela, the **D.501** appeared; this was exactly the same as the previous model except for the HS 12 Xcrs engine; it was equipped with a 20-mm canon installed between the two rows of cylinders and firing through the propeller hub. The first D.501 flew for the first time in June 1936 and a total of 157 machines were built, of which fourteen were for export. This variant was only modified once during production; indeed from N°112 onwards the wings were modified to take Darne or MAC machine guns whose magazines were slightly more voluminous and therefore needed bulges on the upper wing surfaces. The D.503 with a smaller wing, single member undercarriage and engine rated at 690 bhp, flew for the first time in April 1935 and was a better aircraft than both the D.500 and 501 but it was not produced.

When GC I/3 and II/3 received their first D.501s in 1935 they were the first fighter units in the world to be equipped with canon. Subsequently several other fighter groups were also equipped with this machine (GCs I/1, II/1, I/5 and I/8) so that in 1936, twelve front line escadrilles based in mainland France had a 'mixed' strength of D.500s and D.501s, three flying only D.501s and two others (GC II/1) being in the process of converting to the new **D.510.**

Opposite:
Camouflaging military aircraft only started to be general practice at the time of the Munich crisis, a period when Dewoitine planes formed the backbone of the French fighter strength. *(SHHA)*

Above:
Dewoitine D.500 from GC I/8 at Marignane in 1937. The navy anchor painted on the rudder shows that earlier this unit belonged to the *Aéronavale*. *(SHHA)*

The Dewoitines still in service when WWII started like D.510 n°17 seen here in 1940 were camouflaged and given the markings of military aircraft of the period.
(Coll. J. Delmas)

This plane of which the prototype flew for the first time at Toulouse in July 1936 was in fact an improved version of the D.501. The tail had been enlarged, fuel capacity had been increased, the undercarriage modified and simplified (with cantilever struts). It also had a new engine, a HS 12 Ycrs rated at 860 bhp at altitude driving a three-blade propeller which enabled the aircraft to go faster than 250 mph in level flight. A first order for 25 D. 510s was placed with LeO in May 1935; this was followed by two others up to May 1937. At this time, with the nationalisation of the armament industry after the Front Populaire Government's election victory, the SAF had been incorporated into the SNCAM of which Emile Dewoitine had become a board member. In all 120 D.510s came out of the factory and 30 of them were sold for export to several foreign countries, mainly China.

Because of production delays, the D.510 and consequently its predecessors the D.500 and 501 were all out of date when they reached their units: not to forget that on the other bank of the Rhine a certain Messerschmitt was making its first flights... At the time of the Munich crisis, in 1938, the Dewoitine monoplanes nevertheless formed the backbone of the French fighter strength, representing almost 70% of this strength. It is difficult to imagine what would have happened if they had had to be thrown into the battle... With the arrival of more modern fighters like the MS 406, Curtiss H-75, etc., the units gradually replaced the Dewoitine monoplanes. However although a few days before the declaration of war many of them were in the depots or assigned to training schools, there were 145 still in the front line, especially in GC I/1 whose MB 152s had still not been declared '*bons de guerre*' (operational); in GC III/2 and III/3 created in May 1939; and in several other groups (GC II/1, I/8 and II/3) where they were used to bring the units up to strength. They were also used by the ERC 1/561 and 2/561 based at Rouen-Boos and 1/562 at Lyons-Bron. These machines were withdrawn progressively and replaced by the MB 151/152 so that during the winter of 1939-1940 only two defensive patrols made up of Polish pilots still used this type. Only forty or so aircraft of all three versions were inventoried after the armistice. Overseas, the Dewoitine fighters with fixed undercarriage lasted a little longer. In November 1939, ERC 571 at Casablance, Morocco received its D.510s from GC III/1 which had been sent to mainland France with its new MS 406s. The situation was not very good however since the machines were only armed with their machine guns, the cannon not being installed. This regional escadrille merged with ERC 573 itself equipped with the same machines in May 1940 to form GC III/4, a group which was disbanded at the end of August 1940 after being briefly reactivated during the attack on Mers-el-Kebir. It was in French West Africa that the Dewoitines were kept longest in the front line since the *Escadrille Autonome N°6* based at Dakar, became GC I/6 after the Armistice and used them until it was disbanded at the end of 1941 when the Curtiss H-75s arrived in Senegal.

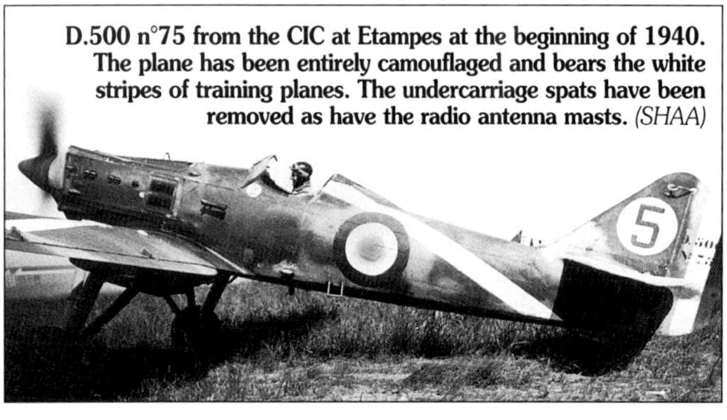

D.500 n°75 from the CIC at Etampes at the beginning of 1940. The plane has been entirely camouflaged and bears the white stripes of training planes. The undercarriage spats have been removed as have the radio antenna masts. *(SHAA)*

TECHNICAL SPECIFICATIONS D.501 (D.510)

Type
single-seat fighter (C1)

Power plants
One Hispano-Suiza 12 Xcrs (12 Ycrs) rated at 660 bhp (680 bhp) on take-off and driving a 10 ft 11 in two-blade Levasseur propeller (10 ft 3 in three-blade Ratier propeller.

Dimensions
Length: *24 ft 11 in (26 ft 3 in).*
Height: *11 ft 3 in (11 ft 11 in).*
Wingspan: *42 ft 6 in.*
Wing Area: *177.589 sq ft.*

Unladen Weight: *2 963 lb (3 289 lb).*
Max. Weight: *4 400 lb (4 246 lb).*

Performance
Max. Speed: *225 mph at 15 510 ft (251 mph at 16 005 ft).*
Range: *537 miles (375 miles).*
Service Ceiling: *26 400 ft (34 650 ft).*

Armament
One 20-mm Hispano-Suiza HS9 canon firing through the propeller hub. Two wing-mounted 1934-model MAC machine guns.

Dewoitine D.510

Dewoitine D.510 n°261 from the *2e Escadrille* of GC I/8. Hyères, France, September 1939. When the call-up came at the end of August 1939, this squadron was not completely equipped with Bloch MB 152s even though this had been planned, and it was obliged to hold on to its D.510s for a few weeks longer in order to keep the unit up to strength.

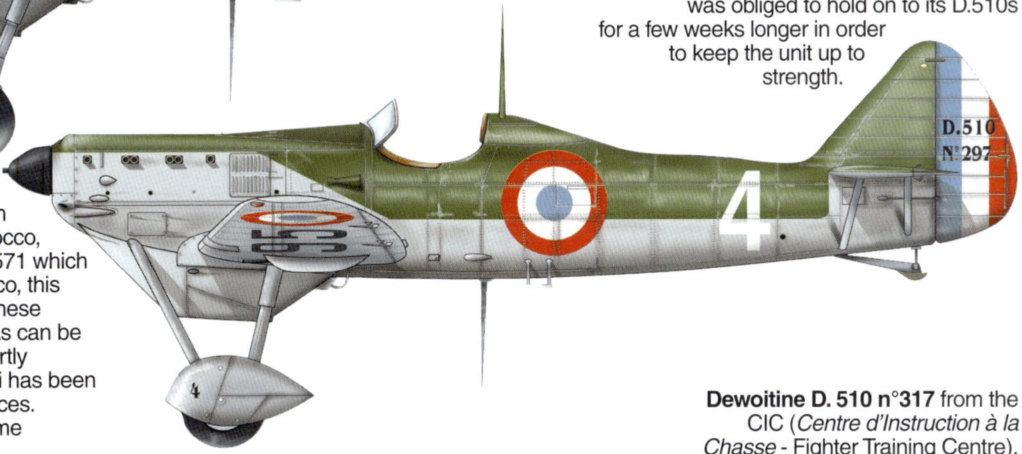

Dewoitine D.510 n°297 from ERC 573. Casablanca, Morocco, May 1940. Along with ERC 571 which was also stationed in Morocco, this *Escadrille* still had some of these fighters in May-June 1940. As can be seen this plane has been partly camouflaged: a coat of khaki has been painted over the upper surfaces. In June 1940, this unit became the *6e Escadrille* of GC III/4.

Dewoitine D. 510 n°317 from the CIC (*Centre d'Instruction à la Chasse* - Fighter Training Centre). Montpellier-Fréjorgues, France, June 1940. This plane which has kept its wheel spats has been camouflaged, partly covering the individual number on the tail in the process; the diameter of the fuselage roundel, which replaced the insignia in the same position, has been reduced in size.

Dewoitine D.510 n°75 from the CIC at Etampes. France, June 1940. The white diagonal stripes which were specific to training aircraft were also painted on the upper surfaces of the wings of this completely camouflaged plane. Here the undercarriage has had its spats removed, a common practice to prevent mud accumulating inside them.

Dewoitine D.520

Above:
Although this photograph is well-known, it does have the merit of being in colour. It is part of a series taken by a German officer while GC III/6 was en route for Syria at Athens-Eleusis airport in May 1941. Here the Groupe's D.520s can be seen sporting the special campaign markings for the Levant and in particular the yellow tails. *(Coll. P. de Gméline)*

The Dewoitine D.520 was without doubt the best fighter available to the Armée de l'Air in 1940. It grew out of one of those coincidences of which only History has the secret. Emile Dewoitine, who was already known for his fighters (D.500 and variants, cf. preceding pages), was anxious to remain one of the main suppliers of single-seat fighters. He was worried that the inevitable nationalisations following on the Front Populaire's election victory would hamper his freedom of action and had in fact decided to go it alone; he had therefore left the SAF and set up a design team with two other engineers, Robert Castello and Jacques Henrat who left with him. They launched themselves into the '500 kph fighter' (312 mph) venture by studying a resolutely modern design built around the Hispano-Suiza 12 Y 21 engine.

For its part, the French General Staff, impressed by the British prototypes of the period (the Hawker Hurricane and especially the Supermarine Spitfire) and the host of orders which had been placed by the British government, had finally brought out the C1 fighter programme in June 1936, which aimed at obtaining an aircraft capable of reaching 312 mph (500 kph) at 13 200 ft, armed with at least one canon and two machine guns, or two cannon.

The initial official programme was very quickly modified, the specified speed rising to 520 kph (325 mph) so that when the project came to fruition, two or three years thence, the planes entering service would not already be out of date as had so often been the case beforehand... Dewoitine's preliminary study which was presented just after the adoption of the speed amendment was refused by the authorities because the plane could indeed not reach the new speed as required. A second project, christened D.520 in reference to this speed, was submitted to the authorities, the machine having only undergone minor modifications. It was accepted this time on 12 January 1937 under the name 'Technical programme A23'. Three other projects responding to this same programme were also chosen: the MS 450, the Loire Nieuport 60 (the future CAO 200, cf. Vol. 1 p.80) and the Caudron-Renault 780, with the Bloch MB 151/152 and Arsenal VG 33 being added to the same programme later.

Nationalised in March 1937, the Dewoitine factories became the Société Nationale des Construction Aéronautiques du Midi (SNCAM) and were entrusted with a lot of sub-contracting work which considerably delayed the project's finalisation.

Two prototypes were ordered officially on 3 April 1938 but the first (N°01) only flew for the first time at Francazal at the hands of the famous Marcel Doret on 2 October of the same year. It was powered by an HS 12 Y 21 rated at 890 bhp on take-off driving a wooden two-blade propeller. This was very quickly changed (leading edge slats removed and three-blade propeller fitted) but it was forced to make an emergency belly landing the following November which delayed the trials even more. Moreover during the trials the engineers noticed that the half-radiators placed in the wings reduced its overall performance.

It was at the end of January 1939 that the second prototype (N°02) took to the air for the first time. It had also undergone several modifications: its ventral radiator had been enlarged so that the elements installed in the wings could be removed; the tail had also been enlarged and the canopy was now a sliding one and not one which swung over to one side. Lastly, the armament had been installed. In this configuration, it reached 515 mph in a dive on 8 February 1939. It was then sent to the CEMA at Villacoublay to continue its trials. It was rapidly joined by the third prototype (N°03) whose tail was slightly different and which was fitted with an HS 12 Y 31 engine. It took off for the first time on 5 May 1939.

D.520 N°48 from GC I/3. Three machines coded I, II and III, were delivered to this unit for its endurance and range trials. Cannes-Mandelieu, April-May 1940. *(Coll. J. Delmas)*

Above: **The SNCAM factory at St-Martin-du-Touch in May 1942, at the time when the building of the D. 520 had been allowed to be resumed. The planes came off the production lines bearing the Vichy Air Force markings, and the camouflage scheme with the wavy delimitation was typical of this period.** *(Coll. J. Delmas)*

Above, right:
The D.520's instrument panel. (Coll. J. Delmas)

D. 520 from the 1re *Escadrille* of GC I/3 in 1942. The plane has been christened '*Limousin*', painted under the windshield. *(Coll. J. Delmas)*

In spite of a difficult start and a certain lack of interest on the part of the authorities, the D. 520 finally benefited from the delays of its direct competitors. When it finished its trials at the CEMA, the VG 33 (cf. Vol. 1, p.26) and the MS 450 had not yet flown, the CAO 200 was only just starting its trials, the MB 151/152 was turning out to be a disappointment and the Caudron 780 had just simply been abandoned.

Meanwhile a first contract for a production series of 200 machines was signed in March 1939, the planes to be delivered between September and December of the same year, at which time it was planned that 100 would be in service; this was, as it turned out, rather optimistic a forecast. This first order was followed by a second for 510 machines. Since the Munich crisis, the weaknesses in the French fighter disposition had indeed been brought to light and had caused the programmes to be modified in the context of the 'reinforced Plan V' of June 1939: 2 623 aircraft of the C1 category were planned of which 710 were D. 520s - to be supplied before the end of 1940.

The building of the D. 520 was designed to reduce production costs and the weight of the machine to a strict minimum. On top of this, the production series aircraft differed from the prototypes especially by their engine, a Hispano-Suiza 12 Y 45 fitted with a supercharger and rated at 945 bhp on take-off. The cockpit had been modified to increase visibility, the armament had been improved and a tail wheel fitted in place of the original skid.

Although the first aircraft of the production series (N°1) did not incorporate all the improvements (it is in fact N°2 which should be considered as the first 'true' production series machine) and flew for the first time on 31 October 1939, production was very far from reaching the figures originally intended (50 per month), slowing down as it did because of difficulties in getting the armament system operating correctly and in stopping the engine from overheating.

The first examples of the D. 520 were delivered to the CEV at Villacoublay in January 1940 and an experimental escadrille was set up with personnel from GC I/3, given the task of eliminating all its youthful defects and establishing which improvements had to be made in order for the machine to be '*bons de guerre*' (operational). Although by 1 April 1940, 139 machines had been built, only 32 had been taken on by EAA 301. On 10 May, the date of the German attack, these two figures had risen to 228 and 75 respectively. At the time the SNCAM produced 10 aircraft per month, a rhythm which again increased the following June so that at the moment of the Armistice 330 aircraft had been taken on by the Armée de l'Air and the Aéronavale (which received 26 machines).

GC I/3 which gave up its MS 406s in mid-March 1940 was the only squadron equipped with the new fighter to be operational at the time of the German attack on 10 May. On this same day, GC II/3 and II/7 were in the process of being re-equipped. Coming from Cannes they were based at Wez-Thuizy and were engaged after 13 May, only to fall back to the south after 10 June and rally French North Africa a week later. This transfer, like that of the other squadrons, is explained by the fact that France had decided to get as many examples as possible of its best fighter away from the enemy by sending them overseas. Definitively settled in Tunisia at the beginning of July, the *Groupe* was finally disbanded on 30 November 1941 as a reprisal for the number of its pilots who had deserted the previous month. This measure was only symbolic since the unit was reformed later as GC III/3.

Above: **A gathering of D. 520s, perhaps GC I/3 in France, or just after they arrived in French North Africa. Note the variety of markings. Note also that the area where the roundels were usually placed on the fuselage and the disks on the tail have both been left totally blank, without any insignia or individual numbers.** *(WingMasters Archives)*

Conversion for GC II/3 was to be completed by the end of January 1940. The squadron however was legally equipped with the MS 406 and could not start its training flights on the new machine before April. Conversion took place at Le Luc-en-Provence and the unit was sent to the northern front after having been equipped (20 May) with its 33 machines. It too fell back progressively to the south after 11 June, its planes finally landing at Alger on the 20th of the same month. It settled there from August. GC II/3 took part in the Syrian Campaign on 17 June 1941 but returned in July with only 13 machines, having lost 17 machines and one pilot in the Levant, with three kills and one probable to its credit.

GC II/3 started its training period on 19 April 1940, its first escadrille (5th) being converted at the end of May and the second (6th) at the beginning of June. The whole groupe was gathered at Nonancourt on 9 June but only stayed a few days at the front before falling back to North Africa where it was disbanded on 25 August.

As for GC II/7, it had indeed started to receive its D. 520s for training at the end of April 1940, but it only carried out its first operational missions with the type on 1 June, obtaining its first kills on the 15th of the same month, just before being sent to French North Africa to set itself up at Bizerte in Tunisia in August.

GC III/6 was the last unit (originally flying MS 406s) to use its brand new D.520s in combat. It fell back on Le Luc to start conversion on 1 June; it then was engaged against Italy which had just entered the war on the 13th. Exactly a week later, it was also sent across to the other side of the Mediterranean and set itself up at Alger-Maison-Blanche which it left for a while, going over to Oran (Algeria) in October 1940 before being sent to Syria in June and July 1941, to support GCs II/3 and I/7. During the month that this campaign lasted in the Levant, it lost 20 machines and five pilots against 21 confirmed kills and four probable.

GC II/6 and III/7 were the last to be converted from MS 406s to D.520s before the Armistice and did not have the time to be committed in combat. The former, which was only partly equipped with the new fighter on 20 June, was in the end disbanded on 10 August according to the Armistice convention which stipulated that only fourteen fighter groupes (among which four using D.520 and all based in French North Africa) would be maintained. GC I/6 was nevertheless set up again at the beginning of 1942 to replace I/6 in French West Africa.

Below: **D. 520 N°14 (the number on the tail has been erased by the censor) was one of the machines used for training pilots in GC I/3 and II/3. The exhaust pipes are not those of the production model and the side fairing for the supercharger air intake has not been fitted.** *(IWM)*

TECHNICAL SPECIFICATIONS

Type:
Single seat fighter (C1)
Power plant:
One Hispano-Suiza 12 Y 45 liquid-cooled V-12 engine rated at 935 bhp on take-off.
Dimensions
Length: *28 ft 5 in.*
Height: *8 ft 4 1/2 in.*
Wingspan: *33 ft 7 in.*
Wing Area: *171.885 sq ft.*
Unladen Weight: *4 670 lb.*
Max. Weight: *5 889 lb.*
Performance
Max. Speed: *330 mph at 13 200 ft.*
Range: *623 miles.*
Service Ceiling: *13 200 ft.*
Armament
One 60-round 20-mm Hispano-Suiza HS 404 canon firing through the propeller hub, two 7.5-mm MAC 34 M 39 belt-fed machine guns (675 rounds each).

Above: **D. 520 n° 423 served in GC III/7 before the Armistice and fought in Syria in June 1941 after having been handed over to GC III/6. Although it bears the yellow and red stripes on the tail, there is no white stripe painted on the fuselage on either side of the roundel which itself does not have the regulation white border for that period. There was clearly a big difference between the official texts and their application in the field.**
(Coll. J. Delmas)

Finally, the conversion of GC III/7 initially planned for June 1940, got under way late since the groupe had to relieve GC III/6. It was sent to Marignane where it received a mix of old and new machines for a total strength of 21 D. 520s at the time of the Armistice. It too was disbanded in the following August.

In 1941, the occupying forces allowed production of the D. 520 to resume in exchange for French factories producing several German aircraft types. Until the end of 1943, more than 500 new machines were built and used to modernise the Vichy air force's fighter strength. According to the plans, ten groupes were to be equipped with D. 520s at the beginning of September 1942, 15 in July of the following year and the two last before 1 July 1944. As can be imagined, things turned out quite differently. However, between August and December 1942, 332 planes were effectively built; they were identical to their predecessors on the outside at least, save for some minor details. In this way, several squadrons which had MB 152s were progressively converted: GC I/1 in June 1942, II/1 in July, II/5 and III/9 in August, so that by September the initial targets had almost been reached. The last time the D. 520s in the Armistice air force were seen in combat was during *Operation Torch*, November 1942, in which the groupes based in French North Africa took part briefly but intensely (namely GC II/3, III/3 -ex-I/3- III/6 and II/5).

The career of the D. 520 did not however end with the dissolution of the Vichy air force and even continued under different skies. The Germans when not using them to train their own fighter pilots handed them over to Romania (150), Bulgaria (96) which used them against the American bombers returning from their raids on the Ploesti oilfields, and to Italy (72) which used the machines stationed in the territory under its control. The Free French also used the machine, in North Africa obviously, but also in 1944 during the reduction of the 'Atlantic pockets' (Cherbourg, Brest, St. Nazaire, Royan etc.).

After the war, the surviving machines were converted into dual control trainers, the D.520 indeed remaining in service until 1953 in a non-operational unit, the *Escadrille de Présentation de l'Armée de l'Air* (French air force demonstration flight). Although there were not enough of them at the beginning of the Battle of France (the D.520 represented only an eighth of the fighter strength in service) and therefore were used only for a very brief period by the units which were lucky enough to get them, the tally for this fighter amounts to 108 probable and 39 confirmed kills (of which ten against Me 109s) for the Battle of France only.

The D.520 ace was *Sous-Lieutenant* Madon from GC I/3, the groupe 'top scorer' for the Armée de l'Air with ten kills of which seven were obtained before the Armistice in June 1940, whereas the famous Le Gloan managed to shoot down five enemy in one single sortie (15 June), completing his tally in Syria (seven kills) the following year! There is a story that a D. 520 ace, *Capitaine* Léon Richard got all his kills against… British planes, the last in May 1942.

Racy, very hard on the pilots, more delicate to handle than a Curtiss H.75 or an MS 406, less powerful than its principal rival, the German Bf 109, the Dewoitine D.520 was nonetheless more powerful at altitude and especially in a dive, where it clearly outclassed its opponents.

Opposite: **A D. 520 from GC III/6 at Algiers in the spring of 1942. The fuselage roundel was no longer worn at the time.**
(Coll. J. Delmas)

Dewoitine D.520

Dewoitine D.520 N°79. Salon de Provence, France, 1940. Fifteen or so instructors from the EMP (*Ecole des Moniteurs de Pilotage* - Pilot Monitor School) were given some D. 520s so that they could carry out their conversion on the spot. In the following June, this plane was assigned to GC I/3.

Dewoitine D.520 N°31 from the CEMA. Cazeaux, France, beginning of 1940. This plane was used for armament trials, which explains why there is a white letter 'A' painted on the tail. It was handed over to the SNCAM after the Armistice before going back into service in GC 1/2.

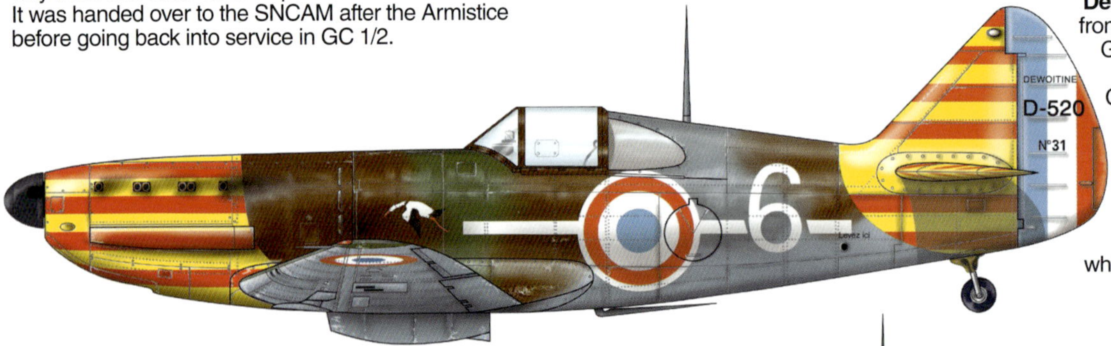

Dewoitine D.520 N° 31 from the *1re Escadrille* of GC I/2. Istres, France, end of 1941. Christened '*Chatelia*', this plane, the same as the previous one, became *Sergent* Lombaert's mount. The planes in the *1re Escadrille* bore a white individual number on the fuselage.

Dewoitine D.520 N°148 from the *1re Escadrille* of GC 1/2. Istres, France, 1941. Disbanded after the Armistice, this famous squadron was re-created in July 1941 at Chateauroux in order to reinforce the French fighter contingent in French North Africa. Following a number of delays, it only went over to Algeria at the time of Operation 'Torch' and finally took up the fight alongside the Allies as GC I/2 '*Cigognes*' in the recently re-created Armée de l'Air.

N.B. in order to help in the plate lay out on this and the following pages, the ventral antenna has been shown folded when it was usually extended in flight.

Dewoitine D.520

Dewoitine D.520 N°73 from the *2e Escadrille* of GC I/3. Suippes, Wez-Thuist, 1940. Nick-named '*Loup-Garou*' (were-wolf), this plane bears the initials of its pilot, *Adjudant* André Carrié, in front of the windshield. He was killed in it on 14 May 1940. This plane was one of the first aircraft delivered to this unit, the first to be equipped with the new fighter in March 1940.

Dewoitine D.520 N°90 from the *1re Escadrille* of GC I/3. Meaux-Esbly, France, end of May 1940. This plane was one of those used by *Sous-Lieutenant* Michel Madon, the ace of aces on D. 520 during the Battle of France with seven confirmed kills and three probable. The planes in the *1re Escadrille* had a blue figure on the tail, with the outline sometimes shaded as here.

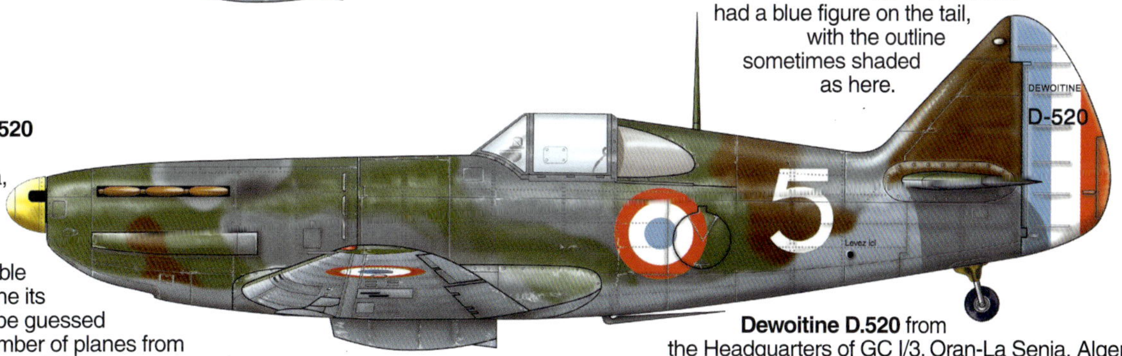

Dewoitine D.520 from GC I/3. Oran-la Senia, Algeria, end of June 1940. As no insignia is visible on this machine its unit can only be guessed at, since a number of planes from various different origins were gathered together in the same place just after their arrival in French North Africa.

Dewoitine D.520 from the Headquarters of GC I/3. Oran-La Senia, Algeria, 1941. This group, placed second in the order of the number of kills obtained by French fighters during the Battle of France, just behind GC I/5, was disbanded in October 1941 after several pilots defected, then re-created a short while afterwards as GC II/3. This plane bears a blazon on the tail containing the two insignia of the *Escadrilles* making up the *Groupe* (SPA 88 and SPA 69).

Dewoitine D.520

Dewoitine D.520 N°48
from GC I/3. Cannes-Mandelieu, France, April 1940. Three unarmed planes numbered with Roman figures were delivered to this group to carry out armament trials. These machines were delivered to the CEMA after 10 May, N°48 being used afterwards in the patrol defending the Dewoitine factory at Blagnac.

Dewoitine D.520 N° 105
from the *2e Escadrille* of GC I/3. Oran la Sénia (Algeria). July 1940. This plane was flown by *Capitaine* Bernard Challes, commanding officer of the *2e Escadrille*. Planes in this flight bore their individual number (white or red) in a white disk painted on the tail.

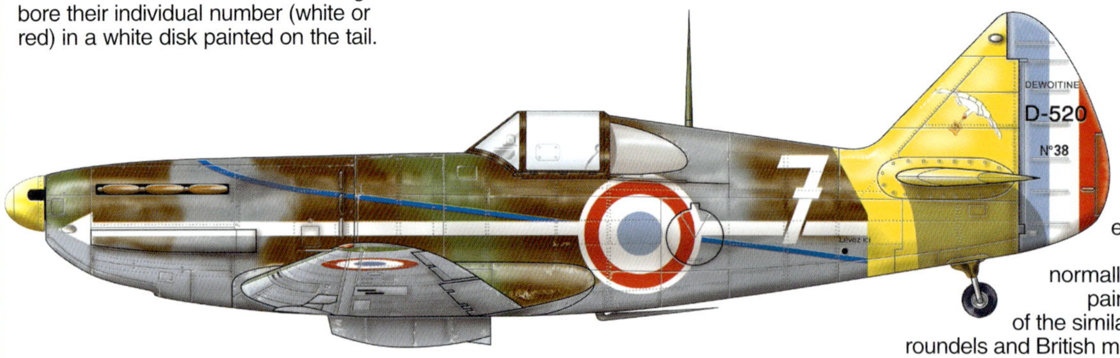

Dewoitine D.520 N°38
from GC II/3. Alep-Nerab, Syria, end of June 1941. Pilot: *Adjudant-Chef* Emile Leblanc. An instruction dated 11 June 1941 stipulated that French planes engaged in Syria had to have their tail and normally the engine cowling painted yellow, because of the similarity between French roundels and British markings.

Dewoitine D.520
from the *4^e Escadrille* of GC II/3. Algiers-Maison-Blanche, Algeria, end of June 1940. The planes in this *Escadrille* were numbered from 21 to 40, the figures being painted red or yellow, as here.

Dewoitine D.520

Dewoitine D.520
from the 4[e] *Escadrille* of GC II/3. Rayak, Lebanon, June 1941. Pilot: *Adjudant-Chef* Bouton. The coloured diagonal stripe indicated which *escadrille* the plane belonged to: blue for the 3[e], red for the 4[e]. Based in North Africa, this squadron took part in the Syrian Campaign in June and July 1941 from where it returned with only 13 planes.

Dewoitine D.520 N°471
from the 4[e] *Escadrille* of GC II/3. Algiers-Maison-Blanche, Algeria, June 1942. After December 1941, the roundel and the fuselage stripe were removed and diagonal tricolour stripes were added on the wings.

Dewoitine D.520 N°49
from the 2e *Escadrille* of GC II/6. Thiès, Senegal, August 1941. This group which had been disbanded after the Armistice was reformed in mainland France in January 1942 and sent to French West Africa a few weeks later in order to replace GC I/6.

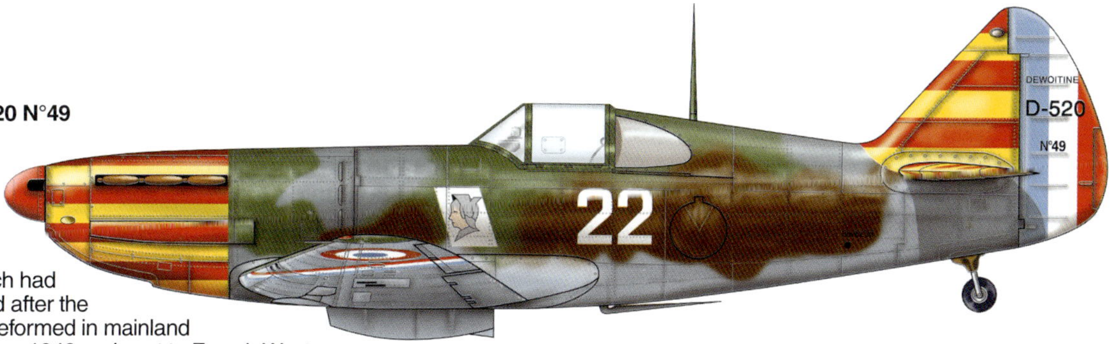

Dewoitine D.520 N°314 from the 5[e] *Escadrille* of GC III/6. Algiers-Maison-Blanche, Algeria, end of 1940. The serial number has not been painted on the rudder flag but painted in white on the fixed part of the tail fin.

Dewoitine D.520

Dewoitine D.520 from GC III/6. Casablanca, Morocco, October 1940. Nick-named '*Ouah Ouah*' (this name was painted on the right-hand side only just below the canopy) this plane was flown by *Sergent-Chef* Pimont. Normally based at Algiers, this group was stationed briefly in Morocco, from October 1940 to January 1941.

Dewoitine D.520 N°314 from the *5e Escadrille* of GC III/6. Algiers-Maison-Blanche, Algeria, beginning of 1941. Pilot: *Lieutenant* Sautier. This is the same plane as on the previous page, to which the unit insignia and the individual number have been added. This machine was destroyed on the ground in Syria on 25 June 1941.

Dewoitine D.520 from the *5e Escadrille* of GC III/6. Algiers-Maison-Blanche, Algeria, July 1940. This squadron took part in the Syrian Campaign, sent out to support GC II/3 and II/7.

Dewoitine D.520 from the *5e Escadrille* of GC III/6. Casablanca, Morocco, end of 1940. This group numbered in its ranks the famous Pierre le Gloan, who obtained 14 of his 18 kills aboard a D. 520.

Dewoitine D.520

Dewoitine D.520 N°357
from the
5e Escadrille
of GC III/6.
Rayak, the
Lebanon, end of
June 1941. Pilot:
Lieutenant Legrand.
As the photographs of these
planes taken during their transfer
to the Lebanon prove, the yellow recognition
markings had been painted on the planes
before they left French North Africa.

Dewoitine D.520 N°132
from the
Headquarters of
GC III/6. Algiers-
Maison-Blanche,
Algeria,
October 1940.
This plane
belongs to
Commandant
Pierre Stehlin, the
squadron's
commanding officer,
whose function was shown
by the letter 'A' on the fuselage and the right
upper wing surface, as well as by the blazon
on the tail fin.

Dewoitine D.520 N°133
from the
6e Escadrille
of GC III/6.
Algiers-Maison-
Blanche, Algeria,
September 1940.
This is the same
machine as the preceding
one whose markings have been
removed, revealed by the way the area behind the
fuselage roundel has been touched up so that the new
individual number can be painted there.

Dewoitine D.520 N°315
from the *6e Escadrille*
of GC III/6. Algiers-
Maison-Blanche,
Algeria, July 1940.
The serial
number has
been painted in
an unusual way
at the base of the
rudder whereas the
rear part of the white
fuselage stripe is wider.

Dewoitine D.520

Dewoitine D.520 N°190 from the *6e Escadrille* of GC III/6. Alger-Maison-Blanche, Algeria, beginning of 1942. This plane is one of the 'escapees' of the group, returning from the Syrian Campaign in the summer of 1941.

Dewoitine D.520 N°273 from GC II/7. Meaux-Esbly, France, June 1940. This plane flown by *Adjudant* Denis Ponteins was lost on 5 June during a dogfight during which another D.520 in the squadron managed to shoot down the famous Werner Mölders. The oil radiator air intake decorated with 'shark's teeth' is unique on this type of plane.

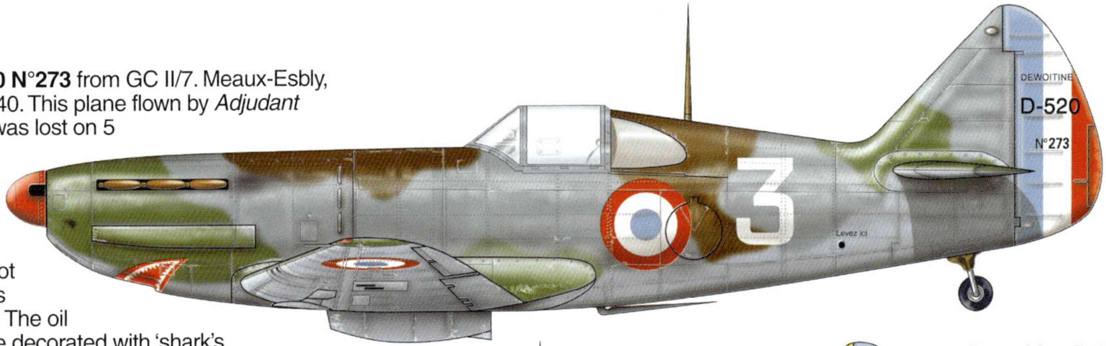

Dewoitine D.520 N°136 from the *3e Escadrille* of GC II/7. Sidi Ahmed, Tunisia. The *3e Escadrille* used letters instead of individual numbers after the Armistice, so this one bears the initials of the pilot's name, *Sous-Lieutenant* Georges Valentin.

Dewoitine D.520 N°347 from the *3e Escadrille* of GC II/7. Sidi Ahmed, Tunisia. This plane was flown by *Lieutenant* Gabriel Gauthier. The letter 'G' had already been attributed within the escadrille to *Lieutenant* Gruyelle so this plane bears two letters in quite an unusual manner on the fuselage to differentiate it.

Dewoitine D.520

Dewoitine D.520 N°245
from the *3e Escadrille* of GC II/7. Sidi Ahmed, Tunisia, June 1941. The recognition markings on the engine cowling have here been interpreted in a rather 'artistic' way... This plane was attributed to *Sergent-Chef* René Martin, an ace with nine kills, which explains the 'M' painted on the fuselage.

Dewoitine D.520 N°346
from the *3e Escadrille* of GC II/7. Bizerte, Tunisia, 1942. Following the removal of fuselage roundel and stripe, the camouflage scheme of this plane, piloted by *Capitaine* Papin (the letter 'P') has been touched up quite a lot in the same area.

Dewoitine D.520 N°248
from the *4e Escadrille* of GC II/7. Sidi Ahmed, Tunisia, March 1941. The way in which the escadrille's insignia (the black panther of SPA 78) is worn, is typical of this escadrille and survived even when the fuselage stripe and roundel were removed. This *Groupe* only received its D.520s in June 1940 and used them against Italy which had just entered the war, until 20 June when they were transferred to Algeria.

Dewoitine D.520
in the patrol defending the SNCAM. Toulouse-Blagnac, June 1940. This unit, made up of three D.520s (N° 312,314 and 321) and civilian pilots from the factory, was originally supposed to protect St-Martin-du-Touch from 10 June onwards but the fighters were attributed to GC III/6 in the end.

Douglas DB-7

Above: **Douglas DB-7 N°12 from the 2^e^ *Escadrille* of GB II/19 over French North Africa in 1940. The white stripe imposed by the Armistice convention runs along the full length of the fuselage.** *(Coll. J. Delmas)*

Although it crashed during its trials, the Douglas Model 7B sufficiently impressed the French purchasing delegation, on a visit to the United States to buy machines which the national aircraft industry could not make, for them to place a first order for 10 machines on 15 February 1939. France thus became the first client for this twin-engined bomber, the USAAC (the future USAAF) not having yet decided whether to buy this plane whose design went back to a 1936 project. Indeed at the time, a team led by the engineer Ed Heinemann - the father of the future A-4 Skyhawk - started to work on an attack bomber whose performances could not fail to interest the American General Staff. The concept of the twin-engine multi-seat aircraft designed specially to carry out this type of mission was in itself a real revolution, as was the choice of tricycle undercarriage and the possibility of changing the nose for a variety of different roles. The prototype flew for the first time at the end of October 1938, but the authorities, against all expectations, were not very interested in the machine, so Douglas decided to export it.

The Armée de l'Air decided to equip itself with this aircraft but with a slightly different version, specially designated DB-7 (for Douglas Bomber Model 7). The first example flew on 17 August 1939 for the first time, powered by Pratt and Witney R-1830-SC3G radials rated at 1 000 bhp each and fitted with single stage superchargers.

In the meantime France had entered the war and the lack of modern planes was increasingly being sorely felt so a second order, for 170 machines this time, was placed in October 1939. These differed slightly from their predecessors since they were powered by P & W S3C-4-G Twin Wasps fitted with two-stage superchargers and fed by Stromberg carburettors. But as the new engines were not available when the first machines were built, Douglas was authorised to continue production of the bomber with the older engines so as not to interrupt the assembly lines, the Twin Wasps being installed from the 131st machine onwards. Other improvements were also made in the course of production like, especially, increased armour to protect the crew and the fitting of self-sealing tanks.

Two other orders were also signed which were not fulfilled owing to the circumstances: they were for 100 DB-7As with Pratt and Witney R-2600-A5B engines rated at 1 600 bhp each and 550 DB-7B-3s (DB-7C according to the builder's coding system) fitted with Wright R-2600-23 engines, to be delivered between October 1940 and October 1941.

All the DB-7Bs (the official French designation) were delivered by boat to Morocco (Casablanca) where they were assembled and painted (they left the factory totally unpainted, without any camouflage but with their nationality markings); the first plane was officially delivered and accepted on 31 October 1939. The Armée de l'Air started to put them into service in January 1940, the assembly line which was entirely new not being operational before that date. Under these circumstances, these bombers were reserved for the *Groupes de Bombardement* (bomber squadrons) based in French North Africa and it was thus that five of them received the new aircraft (GB I/19, II/19, II/61, I/32 and II/32).

On 10 May 1940, only 64 aircraft were actually in service, even though three groups had been completely converted. These units

A Douglas DB-7 at Blida in August 1940. The access trapdoors are open and the forward machine gun ports are clearly visible. *(SHAA 84358)*

DB-7 from the 3^e Escadrille of GB II/61 in flight over French North Africa in 1942. The fuselage roundel is no longer worn. *(SHAA B87 1912)*

were sent to mainland France, GB II/19 being the first to take part in the fighting, when six of its aircraft attacked enemy columns in the St-Quentin region on 31 May 1940. The DB-7s were first 'lent' to the *Groupement de Bombardement N°1* equipped with GM 167Fs (cf. p.30 et seq.) before forming the *Groupement de Bombardement N° 2* on 31 May 1940 when there were enough of them. Deployment of the new bombers in mainland France was far from being straightforward since not all the units had completed their conversion. It was thus that on 5 June GBs I/32, II/32 and GB I/61 were still at Mediouna in French North Africa, whereas GBs I/19 and II/19 (Groupement N°2) were indeed operational at Pithiviers. GB I/61 arrived at Marignane on 9 June but only reached the front two days before the DB-7 units were ordered to fall back... As the Armée de l'Air had chosen to save as many of its more modern aircraft as possible by sending them across the Mediterranean, the four squadrons equipped with DB-7s were re-grouped at Lézignan on 17 June and returned whence they came the following day, after less than three weeks' activity in the best of cases.

Fourteen DB-7s were lost during the Battle of France which was considerable compared with the short length of time they were used. This high attrition rate can no doubt be explained by the type of missions they were sent on, which indeed corresponded to their initial role as assault aircraft, i.e. attacking armoured vehicles or convoys and destroying bridges, but which were carried out in the worst possible conditions imaginable without fighter cover, with heavy Flak everywhere and all the more deadly at low altitudes.

At the time of the Armistice, about half of the first orders had been made in the USA, but France had only received 116 DB-7s (counting the five which were still at sea at the time), of which 39 were in the process of being made or delivered. Taking into account the losses suffered in combat, there were 58 machines still available. 26 machines were delivered and accepted between June and October 1940 whereas out of the 18 still to be assembled, only 12 were in fact finished, the others being used for spare parts. The overall total of DB-7s in the Vichy air force can be put at 96 machines. They all had a lacklustre career since none of the units they equipped took part in any major campaign.

GB I/19 was at Tunis when it was disbanded during the summer of 1940. Its personnel was transferred to GB II/19 which was then renamed ... GB I/19 the following September maintaining its original traditions. The *Groupe* finally set itself up at Blida after having taken part in the raid on Gibraltar in September 1940.

According to the forecasts, GB I/32 should have been converted from MB 200s to DB-7s at the beginning of 1940 but this only happened on 5 July. Based at Casablanca in Morocco it was also involved in the raid against Gibraltar which followed on the heels of the Anglo-Gaullist attack on Dakar (Operation 'Menace' on 22 September). The second *Groupe* of the *32^e Escadre*, GB II/32, settled at Agadir during the summer of 1940 and stayed there until the Vichy air force was disbanded. GB II/61, the last unit equipped with DB-7s, was in the process of being converted to LeO 451s when the Allied landings in North Africa took place on 8 November 1942.

The bombers which survived this particular fighting were used on the Allied side mainly for secondary missions owing to their age; some of them were nevertheless on active service again in 1944 within GB I/31 '*Aunis*' and I/34 '*Béarn*' to reduce the Atlantic pockets.

TECHNICAL SPECIFICATIONS

Type
Twin-engined three-seat light bomber (B3)

Power plants
Two fourteen-cylinder Pratt and Witney R-1830-SC3G (then S3C4-G) air-cooled double-row radial engines rated at 910 bhp each at 11 550 feet.

Dimensions
Length: *47 ft 1 in.*
Height: *17 ft 7 in.*
Wingspan: *61 ft 9 in.*
Wing Area: *463.885 sq ft.*
Unladen Weight: *10 725 lb.*
Max. Weight: *15 125 lb.*

Performance
Max. Speed: *309 mph at 11 550 ft.*
Range: *525 miles.*
Service Ceiling: *25 245 ft.*

Armament
Two 7.5-mm MAC machine guns in the rear gun position. Four forward-firing 7.5-mm Browning machine guns in the nose. 1 760 lb of bombs carried internally.

A line-up of Douglas DB-7s at Blida in August 1940. Note the variety of camouflage schemes. *(SHAA 84353)*

Douglas DB-7

Douglas DB-7 from the 2e *Escadrille* of GB I/19. Blida, Algeria, August 1940. As GB I/19 had been disbanded in September 1940, GB II/19 took up its number and kept its traditions. It was stationed at Blida (Algeria) where it remained until November 1942.

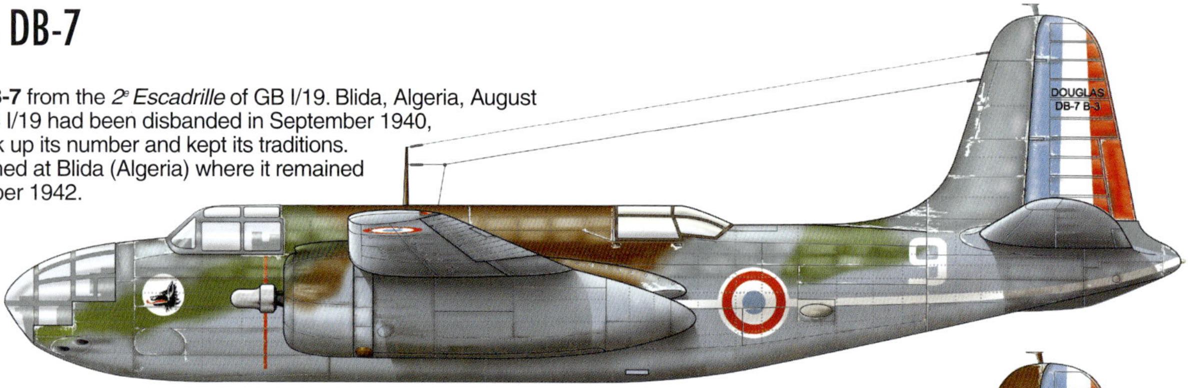

Douglas DB-7 N°4 from GB I/19. Tunis-el Aouina, Tunisia, November 1940. In September 1940, several machines from this squadron took part in a reprisal raid against the British base at Gibraltar.

Douglas DB-7 N°28 from the 3e *Escadrille* of GB II/19. Damville, France, end of May 1940. This unit carried out the first raid by French DB-7s in the Cambrai region on 22 May 1940 only a few days after arriving in mainland France.

Douglas DB-7 N°64 from the 1re *Escadrille* of GB I/19. Blida, Algeria, beginning of 1942. The insignia (SAL 28's elephant) was at the origin that of the 3e *Escadrille* in GB II/19, re-numbered I/19 after the previous one was disbanded in September.

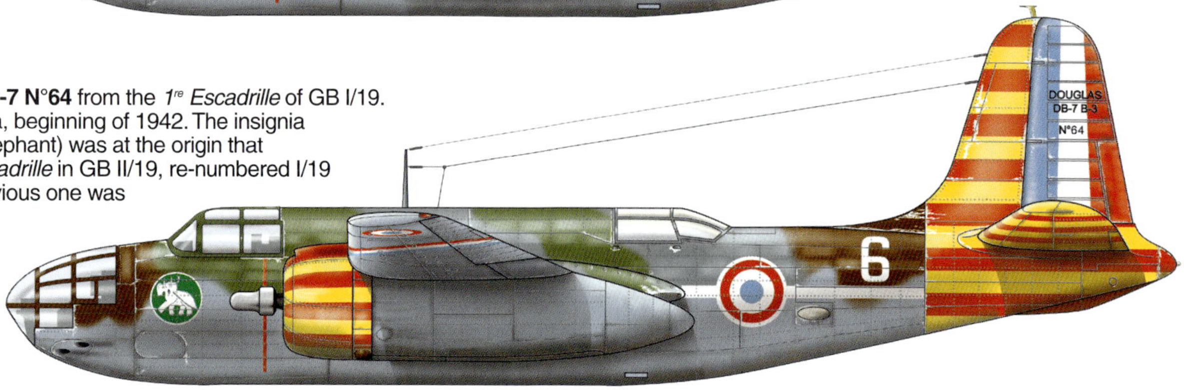

Douglas DB-7

Douglas DB-7 N°45 from GB I/32. Camp Cazès. Casablanca, Morocco, September 1940. According to the original plans, this group was to have been converted to DB-7s at the beginning of 1940, but this was only done on 5 July 1940 and so it did not take part in the Battle of France. This plane belongs to the unit's commanding officer whose usual figure ('13') has been replaced by a zero.

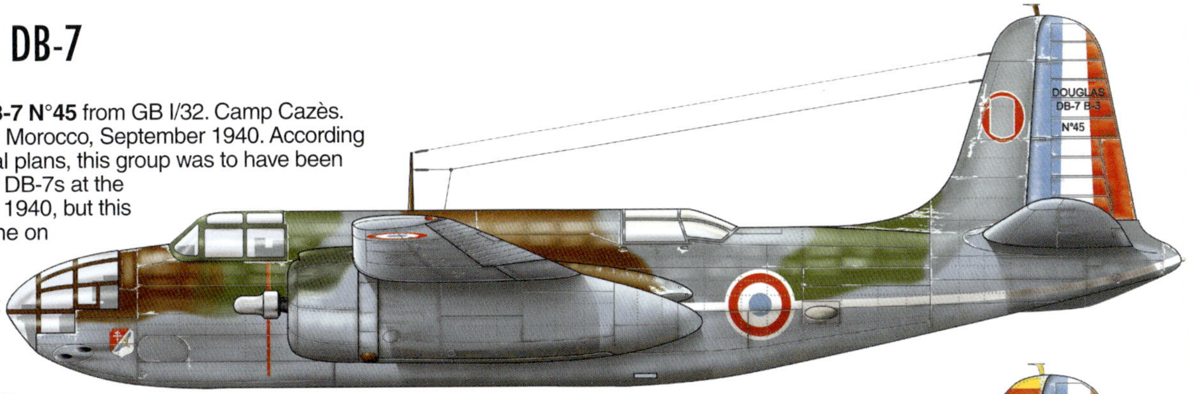

Douglas DB-7 N°91 from the *1re Escadrille* of GB I/32. Casablanca, Morocco, November 1942. This squadron suffered heavy losses on the ground during Operation 'Torch' on 8 November 1942. The Allies wanted to pin these aircraft to the ground so that they would not attack the troops landing. At the time the fuselage roundels had been removed and replaced by oblique tricolour stripes on both faces of the wings.

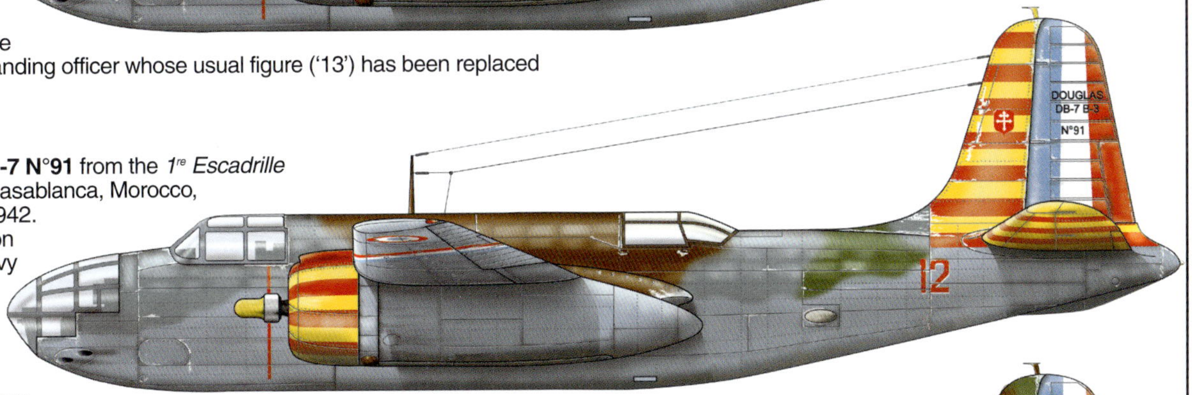

Douglas DB-7 N°79 from GB II/32. Algiers-Maison Blanche, Algeria, June 1940. This group was converted late compared with what was originally planned and did not have time to be sent to mainland France before the Armistice.

Douglas DB-7 N°71 from the *4e Escadrille* of GB II/32. Safi, Morocco, June 1940. The unit insignia were rarely painted on the aircraft before the end of the fighting, but it became more common in the summer of 1940 with the appearance of the Armistice markings.

Douglas DB-7

Douglas DB-7 N°79 from the *3e Escadrille* of GB II/32. Agadir, Morocco, autumn 1940. The two *Groupes* of the *32e Escadre* set themselves up in Morocco after the Armistice, I/32 at Casablanca and II/32 at Agadir. They stayed there until the Allied landings in French North Africa, in November 1942.

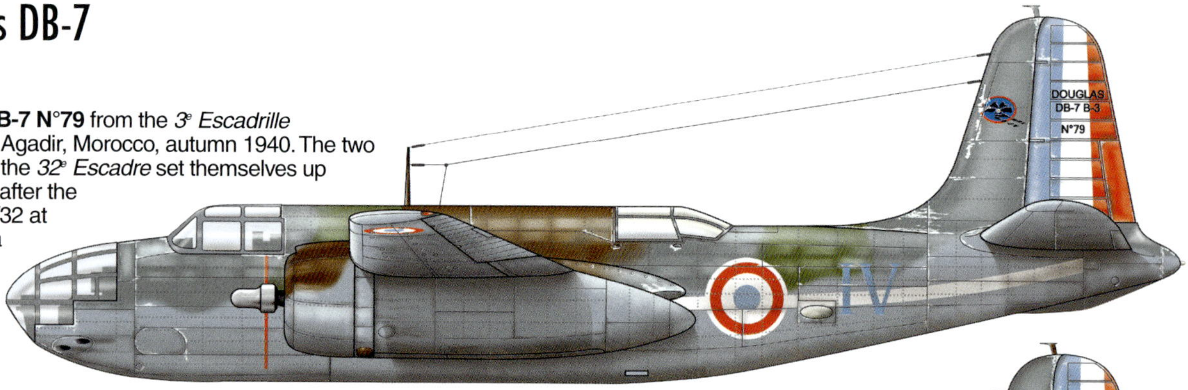

Douglas DB-7 N°2 from the *4e Escadrille* of GB II/32. Agadir, Morocco, autumn 1940. The planes in the *3e Escadrille* bore Roman numerals in place of individual numbers; the *4e Escadrille* used Arab numerals. As was he case on a number of machines after the Armistice, a light shade of camouflage has been applied to this DB-7 because of its frequent flights over the desert.

Douglas DB-7 N°72 from the *4e Escadrille* of GB II/32. Agadir, Morocco, November 1942. This plane was damaged on 9 November 1942, the day following the Allied landings in North Africa.

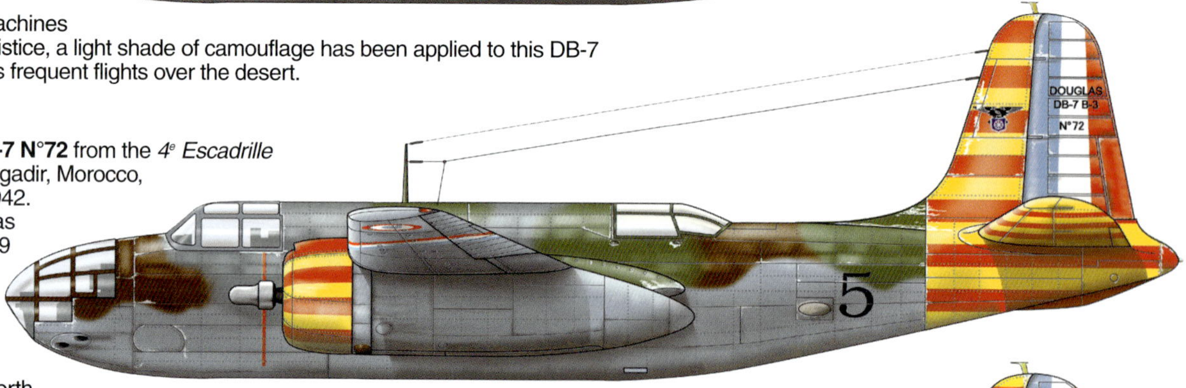

Douglas DB-7 N°50 from the *3e Escadrille* of GB II/61. Blida, Algeria, November 1942. This plane was damaged on the ground by Fleet Air Arm aircraft during Operation 'Torch'. At the time the Groupe was converting to LeO 451s.

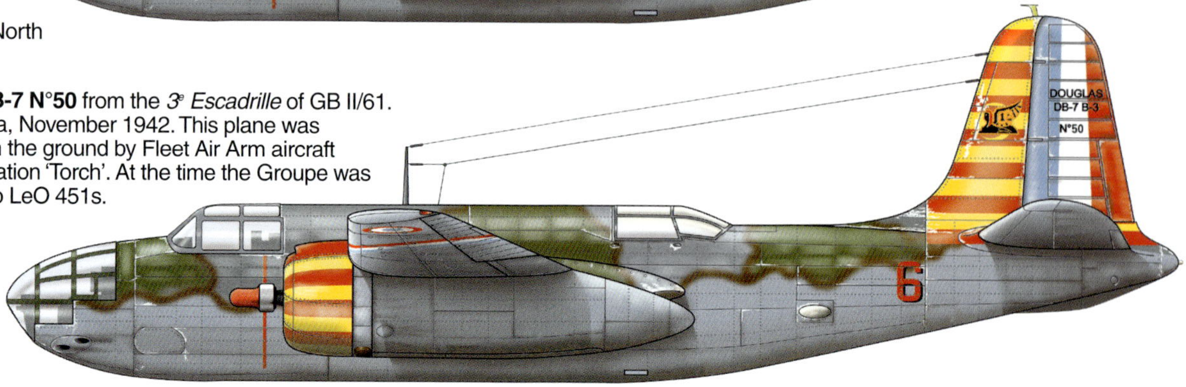

Farman 221, 222 & NC 223

When France entered the war in September 1939, of all the countries at war it was the only one to possess several four-engined heavy bombers in its arsenal. But what seemed to be a great advantage was not in fact one at all, for the design of these machines was very old and therefore totally inadequate for the extreme conditions of this new type of war and as they were only available in small numbers, in the end, they were of no great strategic use, some of them only carrying out a few symbolic operations with panache but which had no effect at all on the outcome of the fighting.

In the context of the 1929 official programme to build a four-engined night bomber (BN4) intended to replace the giant LeO 20 biplanes, the Farman Company which had already successfully built this type of machine (Goliath and Super Goliath), launched itself into the venture with its **F.211** designed by a team led by the Engineer Rocca. On this occasion the biplane formula was abandoned and the plane, fitted with a high wing with struts and a surface area of 1 011.722 sq. ft., was powered by four Farman 9Eb engines each rated at 230 bhp, installed in pairs on stub wings. It was able to carry 1 650 lb of bombs over a distance of 500 miles.

Although this machine which flew for the first time on 19 October 1931 did not give rise to any production series, it was on the other hand at the origin of two other prototypes, ordered officially on 31 January 1930, and answering to the initial programme which had in the meantime had changed to BN5, there being an extra crew member. The first of these prototypes designated Farman F.220 was equipped with four Hispano-Suiza 12 Lbr inline engines rated at 600 bhp each, still installed in pairs and driving four-blade fixed pitch propellers. Thus equipped the new bomber could carry 2.2 tonnes of bombs over 750 miles at a speed slightly over 130 mph. This basic load could be increased to 4.7 tonnes.

Above:
Farman 222 N° 16 (Reg. n° L-158). The two main features of this version are clearly visible from this angle: the lengthened and sloped fuselage front and the outer wing section without dihedral.
(Coll. J. Delmas)

After taking to the air for the first time on 26 May 1932, the machine - the biggest four-engined plane used by the Armée de l'Air at the time (with a 118 ft 8 in wingspan) was finally turned down as a bomber and converted into a long distance mail plane (fuel tanks in the bomb bay, new radiators and propellers) at the request of Air France which put it into service over the South Atlantic after christening it 'Centaure'. It was used by famous pilots such as Guillaumet and Mermoz. This big bird made twenty or so crossings until it was retired in June 1936.

The second prototype, **F.221**, had different power plants - Gnôme & Rhône 14 Kbrs 'Mistral Major' radial engines rated at 730 bhp - as well as a re-designed tail and manually-operated turrets. After flying for the first time during the summer of 1933, it started its official trials the following year during which new, more powerful (800 bhp, supercharged GR 14 Kdrs) engines were installed. On 24 December 1934, the Air Ministry bought fourteen of these bombers, officially designated **F.221 BN5** in the context of 'Plan I' - before the trials had finished. In February 1935, the first two airframes made were attributed to Air France (which received them the following year); a short while later a new amendment to the contract attributed the remaining two planes to the same company. In the company the planes were called **Farman 2200** (N°1 to 4).

The ten remaining F.221s (N°3 to 12, military registrations L-131 to 140) were delivered to the Armée de l'Air between June 1936

and January 1937 who assigned them to the *15e Escadre* based at Avord where they replaced the venerable old LeO 206s.

The career of these four-engined bombers was relatively short because of the appearance of their successors, from the F.222 series. In November 1937, two F.221s (N°4 and 6) were sent to Indochina to serve in the *Escadrille Mixte N°4*, the two extra machines (N° 9 and 10) joining them in 1939. The four machines of this type available in France itself at the beginning of the war were used mainly for filling in the 'gaps' in the *15e Escadre de Bombardement* and only carried out a small number of missions. Having been transferred to French North Africa just before the Armistice, they returned to France at the end of August 1940, GB II/15 being disbanded and its seven F.221s assigned to a transport unit, GT I/15 of which two planes (N° 7 and 8), used during the Syrian campaign in June-July 1941, were finally lost there. As for the four F.221s in the Far East, attached to the Indochina GAA (*Groupe Aérien Autonome* - independent squadron) 2/41, they were used during the war with Thailand in the winter of 1940-41, the two survivors being handed over to Air France after their unit (which had become *Escadrille 2/41*) was disbanded.

The F.221 was made entirely of metal, with a high wing of which the central portion with parallel leading and trailing edges contained the fuel. Two stub wings linked to the main wing by struts supported the coupled engine nacelles, one engine driving a puller propeller and the other a pusher. The fuselage, made up of two main sections joined at the level of the wing trailing edge, held a 16 ft 6 in long bay which could hold a 9 240-lb bomb load; it was divided in two by a gangway giving access to the radio operator's position and the rear turret. Defensive armament consisted of three Farman turrets comprising twin 7.5-mm Lewis guns at first then one single Darne or MAC machine gun.

One of the main defects of the F.221, at least when it came out, was its fixed undercarriage which considerably reduced its performances. As a result, during 1935, the prototype (N°1) was modified and fitted with retractable undercarriage, a modification which meant changing the nacelles and the struts supporting them. The new prototype, **F.222 n° 01**, flew for the first time during the winter of 1935, powered by Gnôme & Rhône 14 Kirs/Kjrs rated at 870 bhp, still fitted in pairs in the nacelles now also housing the forward-retracting landing gear.

Twelve examples of the new plane, designated **F.222.1 BN5** were made by Farman from April 1936 followed by a second series of eight planes, designated **F.222.2,** ordered in October of the same year and which incorporated improvements imposed by the first trials. They also had a different front (three feet longer and with a greater slope) and the wing tips had a pronounced dihedral thereby improving its lateral stability. The 15e Escadre which already had Farman four-engined aircraft of the 'first generation' received the new Farman F.222.1s from April 1937 (GB I/15) whereas F.222.2 was sent to the CEMA in November of the same year to undergo its official trials.

Above:
Mass being celebrated under the wing of Farman 222.2 N° 22 at Oujda, Algeria in 1940. One can but admire the impressive dimensions and the four-blade propellers. The nacelles of this machine which took part in the night-time raid on the BMW factory in Munich on 4 June have been repainted. *(SHAA 801134)*

Meantime, when the aircraft industry was nationalised, Farman was merged with Hanriot within the SNCAC *(Société Nationale de Construction d'Avions du Centre)* which was entrusted with building a third series of 16 Farman F.222.2s (N° 21 to 36) which differed from the previous machines only by their better cooled Gnôme & Rhône 14 N 11 (N 15 at the rear) engines. The last of the thirty-six F.222s left the factory at the end of May 1938, all the planes going to the *15e Escadre* based at Avord. In July 1939, after the *43e Groupe Aérien Autonome* (GAA) was created in French West Africa at Thies in the Senegal, three of these bombers were sent there and carried out maritime surveillance missions in October and November of the same year, chasing U-Boots and the German ' raiders', the pocket battleships *Graf Spee* and *Admiral Scheer.*

In mainland France, once war had been declared, the first four-engined bomber operations over Germany and even

TECHNICAL SPECIFICATIONS F.221 (F.222)

Type
Four-engined five crew night-time bomber (BN5)
Power plants
Four fourteen-cylinder air-cooled Gnôme & Rhône 14 Kbrs (221.1: 14Kdrs. 222.2 14 Kirs rated at 950 bhp) double row radial engines.
Dimensions
Length: *71 ft. (222.1: 71 ft 2 in. 222.2: 71 ft 2 in).*
Height: *17 ft.*
Wingspan: *118 ft 10 in.*
Wing Area: *2 001.918 sq ft.*
Unladen Weight: *19 305 lb (24 200 lb).*
Max. Weight: *33 440 lb (41 140 lb).*
Performance
Max. Speed: *175 mph (222.1: 203 mph. 222.2: 225 mph)*
Cruising Speed: *174 mph.*
Range: *1240 miles.*
Service Ceiling: *26250 ft.*
Armament
Three 7.5 MAC 34 machine guns in two manually operated turrets (front and dorsal) and a ventral bin. 9 240 lb of bombs carried internally.

Czechoslovakia started on 29 December 1939, but were limited to dropping propaganda leaflets and to reconnaissance missions. A few days before the Blitzkrieg, in April 1940, Groupement de Bombardement N°15 was created comprising GBs I/15 and II/15 which took part in several missions over Germany from 14 May (bombing the BMW factories in Munich) but ceased all activities on 15 June.

On 10 May 1940, 28 F.222s were in service in France (11 in GB I/15, 10 in GB II/15 and seven others in various other units, the Aéronavale - *Escadrille 10E* - having received three machines in November 1939). These two squadrons suffered the lowest losses of all the units in the Armée de l'Air, first because of the reliability of the machines themselves and also because their missions were only carried out at night and at high altitudes, but also because of the very few operations carried out due to the limited number of machines available. In all 61 tonnes of bombs were dropped by GB I/15 and 72 by II/15 for the loss of two F.222s, one in an accident on 18 May and the other (N°25) on 30 May on landing, after being damaged during a mission. Most of the Armée de l'Air's four-engined bombers went over first to North Africa then once their groups were disbanded, were re-grouped within GT I/15, a transport unit which gathered together all the Farmans - of all versions - still available.

While in 1938 the F.222 was still being produced, its successor was being designed in the form of the **F.223.** This was a machine which, in spite of its appearance, did not resemble its predecessors. The prototype **(F.223 n°1)** was transformed into a long haul transatlantic carrier at the request of the Air Ministry and, after flying for the first time in June 1937, was entered for the Istres - Damascus - Paris race which turned out to be a fiasco for the French aircraft industry... Powered by Hispano-Suiza 12 Xirs/Xjrs liquid-cooled engines, the only F.223.1 beat the distance record in October 1937

Farman 221 in flight. The fixed undercarriage characteristic of this version can be seen clearly here, like the engine nacelles which have been left unpainted, on the lower wing stubs. The plane is painted all over in a dark brown colour called '*tête de nègre*'.
(Coll V. Gréciet)

then, christened '*Chef-Pilote L. Guerrero*', was also put into service with Air France over the South Atlantic.

The second four-engined aircraft to be built was the true prototype of the new bomber and was designated **NC 223 n°01** since in the meantime, as already mentioned, Farman had been incorporated into the SNCAC of which all the aircraft bore the NC prefix. It was slightly different compared with the long haul version: its fuselage was more refined and it was equipped with three gun positions with turrets. It was now powered by HS 14 Aa 08/09 radials, boosted by superchargers and rated at 1 100 bhp at 13 200 feet. After having flown for the first time on 18 January 1938, it was sent to the CEMA in June for its trials, during which the engines turned out to be unreliable and had to be replaced by HS 12 Y 29 liquid-cooled inlines, rated at 910 bhp at take off which were more reliable and hardier. Thus modified it was given the designation NC 223.3, **NC 223.2** being a prototype fitted with Gnôme-Rhône radial engines which was not produced in any numbers. The official trials were very often delayed and only resumed in May 1939.

In May 1937, 18 F 223 BN5s had been ordered officially; this was reduced to eight machines. The first production series machine (N°2) having taken off for the first time on 10 January 1940 and the last (N°8) on the following 28 March, an extra order for a further 10 machines was placed, reduced later to seven.

The F.223.3 was made entirely of metal with a big wingspan.

This Farman 222.1 N° 10 (L-153) seen at Toulouse at the time of the Armistice is standing alongside a large quantity of materiel of all types (Amiot 143, MB 200, etc.) which can be seen in the background.
(WingMasters Archives)

Above:
Farman 222.1 N°1 from GB I/15 (in the foreground) and 222.2 n°15 from GB II/15 seen at Avord before the war: the differences between the two versions are clearly visible. The *Escadrille* insignia have been painted on the fuselage. *(SHAA 383340)*

The wing was placed high up on the fuselage and held all the fuel. It was made of two spars covered with metal and held by two struts to the stub wings which supported each pair of engines. It now had a twin tail and the crew which was now housed in a square cross-section fuselage divided into three compartments still consisted of five men: a bombardier-navigator serving an MAC 7.5-mm machine gun on a flexible mounting in the nose, the radio operator sitting in the front slightly lower than the pilot's seat. The pilot and the co-pilot sat side by side. Behind them was the bomb bay which could carry 4.2 tonnes of bombs, still divided in two by a gangway giving access to the rear turrets each armed with a 20-mm HS 404 canon.

Out of the fifteen NC 223.3s ordered, including the prototype, only eight had been received before the Armistice, the first two being delivered to GB I/15 on 24 May 1940 which already had four of these bombers when it was sent to French North Africa on the following 15 June. The long delays in delivering these machines were mainly due to delays putting the final touches to the on-board armament. GB II/15 received the same number of aircraft at the beginning of June but did not have the opportunity to use them operationally, conversion not having been completed before the unit was sent over to North Africa. All the NC 223.3s except for N°14 which was damaged at St-Yan, crossed the Mediterranean on 17 and 18 June and set themselves up in Algeria.

After the Armistice, some planes (N° 2.7.8 and 10) with a civilian registration, were handed over to Air France after being converted to carry passengers (eight in place of the bombs), the others being attributed to GB I/15 which had been put back into service in October 1940 and which had recovered the NC 223.3s from GB II/15, itself disbanded in September 1940. Two machines were used for long haul flights between mainland France and the French territories in the Levant (Beirut and then Rayak in the Lebanon) but these lines only lasted a few months. One of the planes, N°2, present in the Levant was recovered by the FAFL during the Syrian Campaign. Its career on the other hand was very long since it was only scrapped in 1945 when it returned to France. When the Unoccupied Zone was invaded by the Germans, they captured two machines and assigned them to the Luftwaffe which put them into service from le Bourget at the beginning of 1943.

TECHNICAL SPECIFICATIONS (NC 223.3)

Type
Four-engined five-crew night-time bomber (BN5)

Power plants
Four 12-cylinder water-cooled Hispano-Suiza 12 Y 29 inline engines rated at 910 bhp each at take-off.

Dimensions
Length: *72 ft 3 in.*
Height: *16 ft 7 1/2 in.*
Wingspan: *110 ft 3 in.*
Wing Area: *1 411.029 sq ft.*
Unladen Weight: *23 177 lb.*
Max. Weight: *42 185 lb.*

Performance
Max. Speed: *218 mph.*
Cruising Speed: *175 mph.*
Range: *1 500 miles.*
Service Ceiling: *26 235 ft.*

Armament
One flexible 7.5-mm MAC machine gun in the nose, one 20-mm HS 404 canon in the rear dorsal turret and another in the ventral turret. 9 207 lb of bombs.

Farman 224 TT N°2 from GT I/15 in transit in Algeria in 1942. Designed as '*Centaure II*' for Air France, the six planes built (F.222 wings, GR 14 Kdrs engines and a fuselage able to carry 40 passengers and freight) were re-named F.224 TT (Troop Transport) when they were handed over to the Armée de l'Air but they did not have the opportunity to carry the *Groupes d'Infanterie de l'Air* (GIA- airborne infantry squadrons) as planned before the Armistice and the surviving planes were used by GT I/15. *(Coll. J. Delmas)*

Farman F. 222.1 & F. 222.2

Farman F.222.1 N°6
(Reg. N° L-148) from the *2e Escadrille* of GB I/15. Avord, France, September 1939. The *15e Escadre* started to receive this type of plane painted '*tête de nègre*' (dark brown), one of the colours used particularly by French night bombers at the time.

Farman F.222.1 N°2
(Reg. N° L-144) from the *3e Escadrille* of GB II/15. Avord, France, September 1939. The twelve machines of this first variant could be distinguished by their shorter, less sloping fuselage nose, their wings without dihedral and their Gnôme-Rhône 14K engines.

Farman F.222.1 N°11
(Reg. N° L-153) from GB II/15. Rheims-Coucy, France, beginning of 1940. The introduction of the roundel on the fuselage of the Farmans meant that the unit insignia were removed to make room for them. On this model the nacelles and the engine cowlings have been painted brown to make them a little less visible.

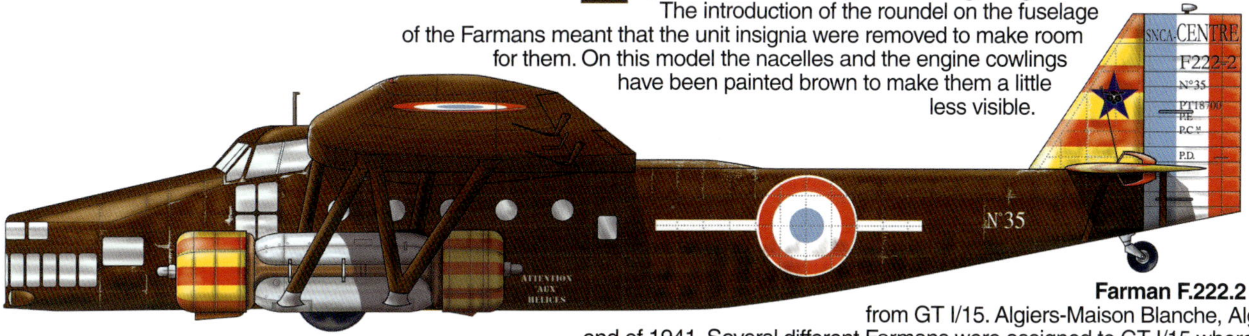

Farman F.222.2 N°35
from GT I/15. Algiers-Maison Blanche, Algeria, end of 1941. Several different Farmans were assigned to GT I/15 where they were used as transport planes after being disarmed (the turrets were removed). The insignia of the *1re Escadrille* of GB I/15 was nevertheless kept on the tail fin, its traditions having been taken up by the new unit. This aircraft was the last F.222 built and was delivered in the spring of 1938. Like all the planes of this type, its front was longer and sloped more; it was powered by Gnôme-Rhône 14N engines and the outer wing sections had a dihedral.

Farman F.222.2 & NC 223

Farman F.222.2 N°22 (Reg. N° L-217) from the 2^e^ *Escadrille* of GB I/15. Rheims, France, April 1940. After carrying out reconnaissance missions and leaflet dropping during the 'Phoney War', this squadron started its bombing operations against Germany on 14 May 1940. On this plane which took part in the raid on the BMW factories in Munich during the night of 3 to 4 June, the insignia and the unit stripe were rubbed off and partly covered by the fuselage roundel and a nickname, '*Pollux*', was painted under the front turret.

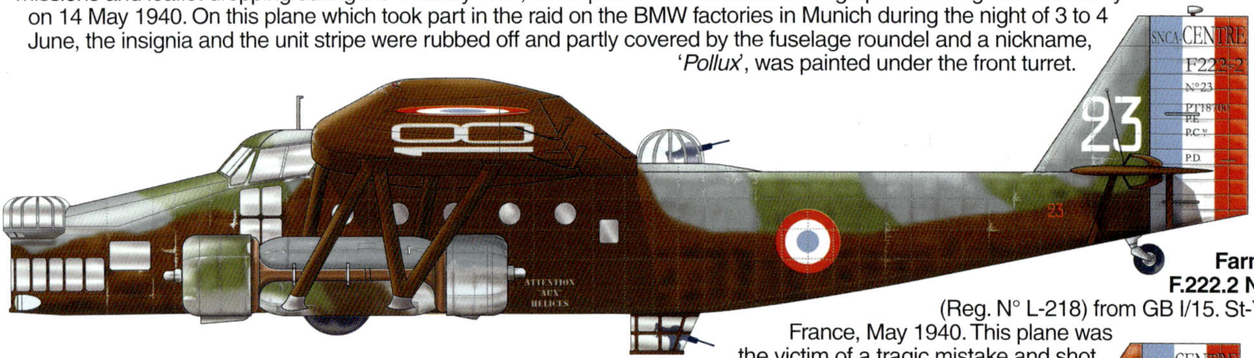

Farman F.222.2 N°23 (Reg. N° L-218) from GB I/15. St-Yan, France, May 1940. This plane was the victim of a tragic mistake and shot down by French anti-aircraft guns on 11 May 1940 as it arrived at St-Yan while the group was being transferred to this airfield. It was one of the rare four-engined machines to have camouflage on the upper surfaces.

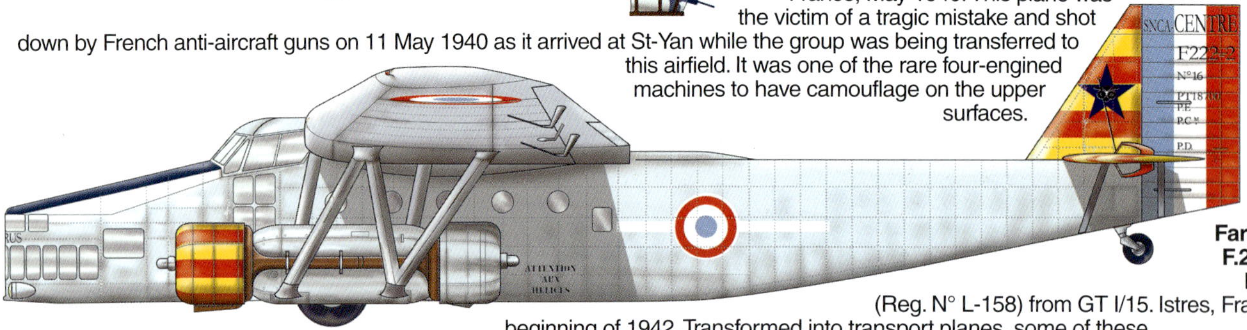

Farman F.222.2 N°16 (Reg. N° L-158) from GT I/15. Istres, France, beginning of 1942. Transformed into transport planes, some of these four-engined machines had their camouflage removed and were left with a bare metal finish or what is more likely, they were painted with aluminium paint. An anti-glare part was painted (black or night blue) in front of the windshield. Note the very small white fuselage stripe.

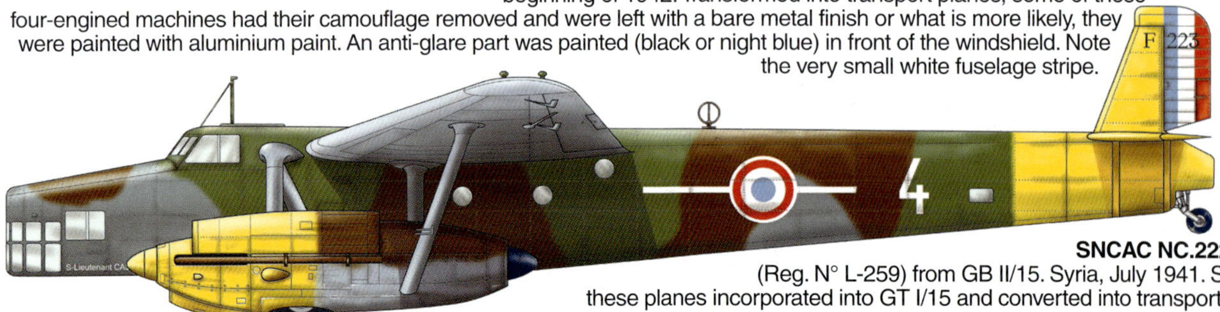

SNCAC NC.222.2 N°4 (Reg. N° L-259) from GB II/15. Syria, July 1941. Some of these planes incorporated into GT I/15 and converted into transport planes after the Armistice took part in the Syrian campaign in June-July transporting men and supplies to the '*Levant*' (French Middle East). One of these machines (N°2) was recovered there and used by the Free French (FAFL).In spite of their designation, these planes were very different from their F.221 and F.222 predecessors and were the last heavy bombers designed by Farman before it was nationalised and incorporated in the SNCAC.

Glenn Martin 167F

At the end of the thirties, two official programmes had been launched on either side of the Atlantic which had the same requirements and which ended up, with circumstances helping a bit, with France buying Glen Martin light bombers.

Indeed, in the USA, the 1937 programme asked for a three-seat twin-engined aircraft flying at least at 200 mph and capable of carrying a bomb load of 1 200 lb over 1 200 miles. Various aircraft manufacturers lined up and presented their projects. Among them, two (Bell 9 and Boeing-Stearman X-100) were not followed up but the three others were to have an excellent future: the North American NA-40, the future B-25 Mitchell, the Douglas Model 7 (see page 20) and the Glenn Martin 167W. The French were greatly interested by these two models, even before they had flown because they corresponded almost exactly to the 1936 A3/B3 programme. This interest was again confirmed with the launching of the new A3 programme whose objective was the acquisition of a three-seat twin-engined reconnaissance aircraft and bomber which had a top speed of at least 295 mph at 13 200 feet, with four hour's endurance and carrying 440 lb of bombs as well as being armed with two machine guns firing forward and two others for defence.

A first purchasing mission to the United States between December 1938 and February 1939 bought a certain number of these twin-engined aircraft which were so cruelly lacking in France's air force: 100 Douglas DB-7s and 115 Glenn Martin 137s to be delivered between 1 July and the end of 1939, and finally 795 Pratt and Witney engines to equip these machines and also the Curtiss H-75s which France had also ordered.

At the same time the prototype of the GM 167 bought by the USAAC and designated XA-22 flew for the first time on 14 March 1939, but the plane was never ordered into production since the US Army Air Force preferred a variant of the DB-7. Some months later, war having started, new orders for 130 GM 137s to be delivered before July 1940, followed by 200 Glenn Martin 187 (the future British 'Baltimore') for delivery in the period October 1940 to October 1941. The first example of the new bomber left the factory in August 1939, but deliveries were delayed by the embargo on the sale of military materiel to belligerent countries decided upon by the USA which was neutral at the time.

Originally, the Martins were to have been shipped to France in crates by boat to le Havre then conveyed to Carpiquet near Caen. Although this is what happened to the first nine machines in December 1939 (before being transferred to French North Africa), it was decided to set up an assembly plant in Casablanca, Morocco, thanks to a subsidiary of the SNCAN. In fact this assembly line only started to operate in January 1940. Assembly of the machines was delayed because certain items of equipment produced in mainland France only reached French North Africa in February...

Originally the Martin 167 A3s, as their suffix indicates, were intended to modernise six of the fifteen reconnaissance groups already on active service. But as they could also take on the role of bomber, they served to convert squadrons based in North Africa, near where they were assembled (GB I/62, II/62, GB I/63 and II/63), just like the DB-7s. Since deliveries had been delayed, the original plans were modified to take into account the delivery of the machines. It was thus that, for example, GB I/39 which originally was to be equipped with DB-7s, was in the end equipped with GM 167s. An experimental flight was set up in Algiers in January 1940 and in the following May, 77 examples of the twin-engined bomber were distributed among the five groupes already mentioned together with GR I/61.

Two initial units (GB I/62 and I/63) were transferred to mainland France just before the beginning of the Battle of France and formed the *Groupement de Bombardement N°1*, assigned to the ZOAN. However, a number of these twin-engined planes was not considered as '*bons de guerre*' (operational), various items of equipment not having been installed (bomb-launchers, machine gun ammunition, etc.). On 13 May it was GB II/63's turn to cross the Mediterranean. As for GB II/62, it was only transferred on 18 May because of a lot of orders and counter-orders caused by uncertainty about Italy's entry into the war delaying its transfer. On 20 May 1940, six light bomber squadrons were available in mainland France (four with GM 167s and two with DB-7s); the French Martin 167s carried out their first mission on 22 May (five machines from GB I/63 sent over the Cambrai region). The last operations took place on 14 June; the order to go over to French North Africa to continue the struggle against Italy from the other side of the Mediterranean was announced the previous day with effect from the 16th, all movements to be completed by the 18th. Several GBs which had

A line up of GM 167Fs from GB I/62 at Meknes at the end of April 1940. The planes bear nationality markings and a red flash on the bare-metal fuselage which they kept briefly after their transfer to mainland France.
(SHAA 972916)

come back from mainland France carried out several missions between 22 and 24 June against Italian targets in Tripolitania, even Sardinia or Sicily. Finally GB I/39, the only unit equipped with GM 167s based in the Levant could only carry out reconnaissance missions with its machines because it only really became operational at the moment that the Armistice was signed with Italy.

When the fighting stopped, 234 Martin 167s had been unloaded in Casablanca, thirty or so were still in their crates, some still at sea, and forty had been handed over to the Aéronavale. As for the fighting in mainland France, it had caused the loss of just over forty planes during the four weeks they were in France (according to the sources, 52 167s had been sent to the front line to which must be added 45 reserve machines based at Marignane and Toulouse. The Armistice did not mean the end of GM 167 assembly at the Casablanca plant and at the end of 1940 almost all the planes had been put together, some of them being used as a source of spare parts.

According to the conditions of the Armistice, no squadron equipped with Glenn Martins was allowed to be stationed in mainland France. Only four bomber squadrons - GB I/62, II/62, I/63 and I/39 - to which must be added some planes from the Aéronavale, were kept overseas, as well as two reconnaissance units - GR I/22 in Morocco and GR I/61 in Algeria. GB I/32 exchanged its Martins for the DB-7s it had left behind when it was transferred, re-equipping being completed at the beginning of July 1940. GR II/36 was to have received GM 167s but it was disbanded at the end of August, before the original plans were put into effect. GR I/61 was originally intended to be a bomber squadron after being equipped with DB-7s. This measure was finally abandoned and the unit remained a reconnaissance unit until the beginning of 1941 when it was disbanded.

Glenn Martin 137 of the SAM, registered as F-BACS and painted entirely yellow, was the first plane to be transformed into a fast long distance courier. It is seen here at Thiès, Senegal in 1941. *(SHAA 832818)*.

GBs I/61 and I/62 later joined by I/63 and II/63, recreated in 1941, were sent to French West Africa where they protected the region and the port at Dakar together with some fighter units, resisting several British attacks (Operation 'Menace' from 24 to 26 September 1941). At the end of July 1940 GB I/22 received a number of GM 167Fs with which it formed a *Groupement de Reconnaissance* the following August comprising GR I/52 and I/35. Based in Morocco, it sent a detachment to Syria in June and July 1941 and received some LeO 451s in the course of 1942 thanks to which it became a bomber squadron, re-equipping being completed the following October.

Finally, GB I/39 was logically engaged in the Syrian campaign in June-July 1941, being brought up to strength when it received the Martins from GR I/22. Having lost several machines during different attacks on its airfields, it left Syria with only 12 out of the original 22 planes before being disbanded on 30 September 1941.

Apart from their military use, several 167Fs were converted into 'long distance liaison' and VIP transport aircraft after the cessation of hostilities. The SSLA *(Section Spéciale de Liaisons Aériennes)*, established in Algiers and flying under the control of the Armistice commission with aircraft bearing special markings and a civilian registration and flown by civilian pilots, used these planes as did the SAM (*Section d'Avions Ministériels* - Ministerial Aircraft Section - a distant ancestor of the GLAM) which was created in mainland France and which was equipped with eight converted GM 167s, among other planes.

A GC I/5 Curtiss H-75 simulates a rear attack on a GM 167F from the GR I/22 in November 1940. *(SHAA B79522)*

TECHNICAL SPECIFICATIONS

Type
Three-seat twin-engined light reconnaissance bomber (A3/B3)

Power plants
Two Pratt and Witney R- 1830-SC-3G rated at 1 065 bhp (max) each and driving an 11 ft 6 in Curtiss fixed-pitch propeller.

Dimensions
***Length:** 46 ft 11 in.*
***Height:** 16 ft 6 in.*
***Wingspan:** 61 ft 7 in.*
***Wing Area:** 512.318 sq ft.*
***Unladen Weight:** 10 758 lb.*
***Max. Weight:** 15 672 lb.*

Performance
***Max. Speed:** 305 mph.*
***Cruising Speed:** 255 mph.*
***Range:** 1 125 to 1 750 miles.*
***Service Ceiling:** 29 139 ft.*

Armament
Four 7.5-mm FN machine guns installed in pairs in the leading edge of each wing. One MAC 34 machine gun in the upper gun position and another in the lower one. Offensive load of 1 760 lb made up of only 22-lb or 110-lb bombs.

Glenn Martin 167F

Glenn Martin 167F N°27 from the *Groupement de Bombardement n°1* (unit unknown) during the Battle of France, May-June, 1940. The Martins were transferred to mainland France even though they were not yet '*bons de guerre*' (operational), and often not even camouflaged; the insignia were rarely worn before the Armistice, making identification very problematic.

Glenn Martin 167F N°35 from GB I/39. Rayak, the Lebanon, June 1940. This was the only unit to be stationed in the 'Levant' in 1940 and equipped with this twin-engined bomber even if only partly. This group only became operational for attacks on possible Italian targets at the time that the Armistice was signed... In the Middle East, the Martins were used in the three-seat observation (A3), bomber (B3) or ground attack (AB3) roles depending on the circumstances. As with the MS 406s in GC I/7 which were also based in the Lebanon, these planes wore an Arabic numeral as their individual number. This machine (N°2) was destroyed in an accident when the squadron moved from Aleppo (Syria) to Rayak on 5 April 1941.

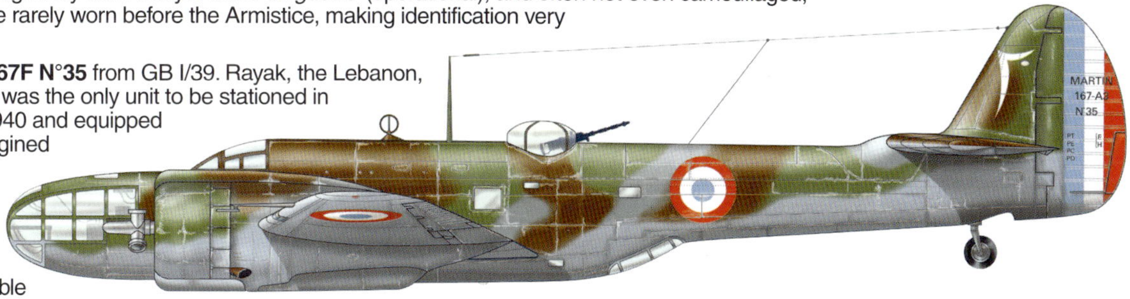

Glenn Martin 167F N°84 from GB I/62. Meknès, Algeria, April 1940. Still bearing this highly coloured livery and without their bomb-launchers, the Martins in this group were transferred to mainland France at the beginning of May 1940 where they were camouflaged. GB I/62, one of the units with the most antiquated equipment (LeO 20B biplanes) was only converted to these modern twin-engined bombers at a late stage. Its re-fitting, planned to be completed by January 1940 had not even been started in May. N°84 had a particularly short life since it was damaged on 25 May 1940, fifteen days after its arrival...

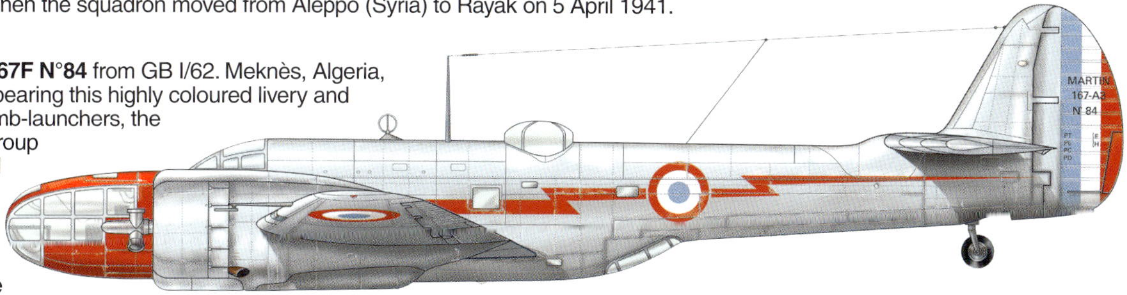

Glenn Martin 167F N°187 from GB I/62. Toulouse, France, 28 June 1940. Apart from its camouflage which was painted on the spot, the Martin 167Fs which were used during the Battle of France very often only bore an individual number enabling them to be identified within the unit only. This squadron, one of the first to be transferred to mainland France at the beginning of May 1940, returned to French North Africa a month later and was then sent to Senegal in French West Africa.

Glenn Martin 167F

Glenn Martin 167F N°57 from the *1re Escadrille* of GB I/62. Bamako, Mali (French West Africa), 1942. Sent first to Thies, Senegal, in mid-July 1940, this unit then left for Bamako in September where it formed the Groupement n°2 for French West Africa with GB II/62. The Martins in this squadron, so the story goes, detached and abandoned at Libreville (Gabon), were recovered by the Free French (FAFL).

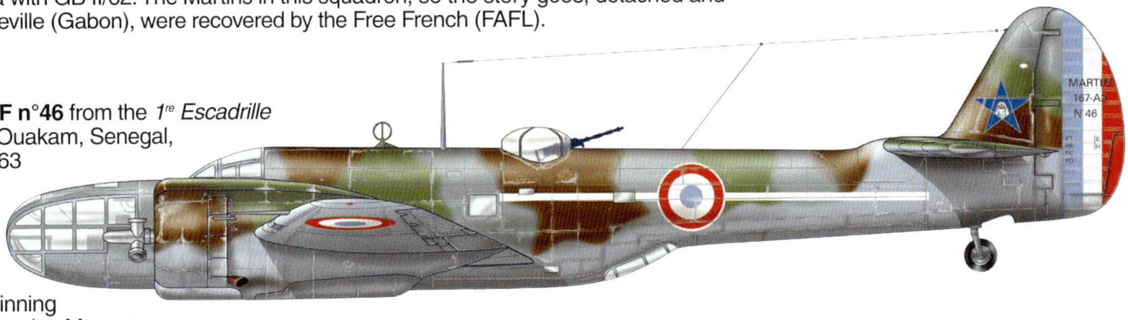

Glenn Martin 167F n°46 from the *1re Escadrille* of GB I/63. Dakar-Ouakam, Senegal, end of 1940. GB I/63 which made up the *Groupement de Bombardement N°2* along with GB I/62 at the beginning of May 1940, returned to Morocco the following June and was then sent to Dakar in August 1940.

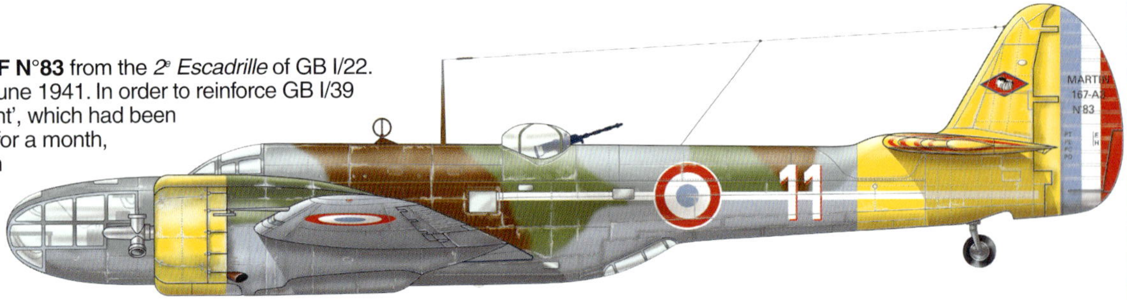

Glenn Martin 167F N°83 from the *2e Escadrille* of GB I/22. Rabat, Morocco, June 1941. In order to reinforce GB I/39 based in the 'Levant', which had been operating by itself for a month, a detachment from GB I/22 was transferred there at the end of June 1941 but only eight of the ten machines sent there actually reached their destination. The special markings for this theatre of operations were painted before the planes left for the Middle East. n°83 was destroyed on the ground at its airfield at Quseyr in a raid by British P-40s on 29 June.

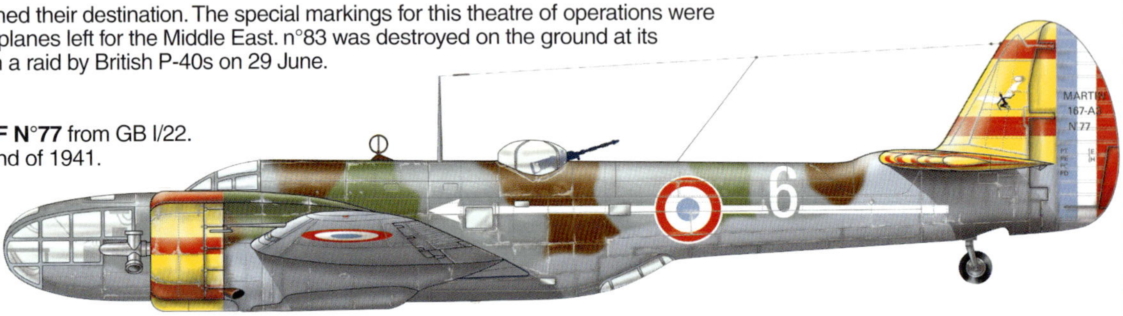

Glenn Martin 167F N°77 from GB I/22. Rabat, Morocco, end of 1941. At the end of July 1940, GR I/22 finally took delivery of its GM 167Fs with which it formed a reconnaissance squadron in August with GRs I/52 and I/35. When the latter was disbanded, GR I/22, now independent again, set itself up at Rabat-Salé, Morocco. The following year it received some LeO 451s whose production had been resumed and was converted into a bomber unit, its re-fitting being finished in October. N°77 survived the Syrian campaign unlike the aircraft above.

Glenn Martin 167F

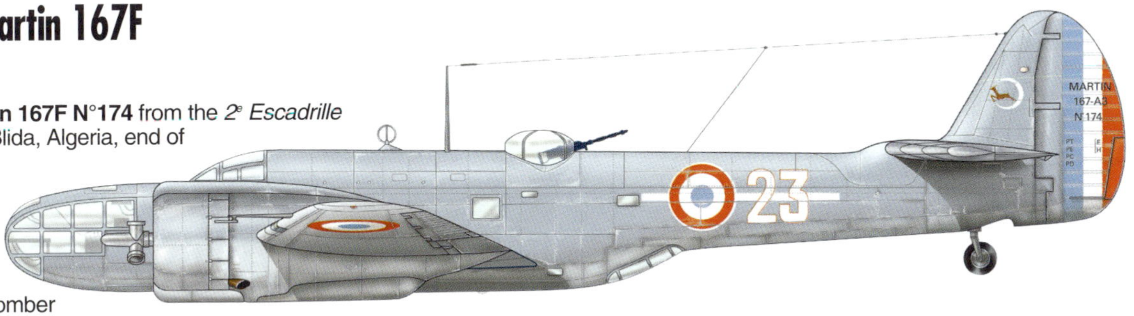

Glenn Martin 167F N°174 from the *2e Escadrille* of GR I/61. Blida, Algeria, end of 1940. It had originally been decided that GR I/61 would become a bomber squadron after being equipped with DB-7s. This idea was finally abandoned and the unit remained in its reconnaissance role until the beginning of 1941 when it was disbanded. Most of its machines were handed over to GB II/63.

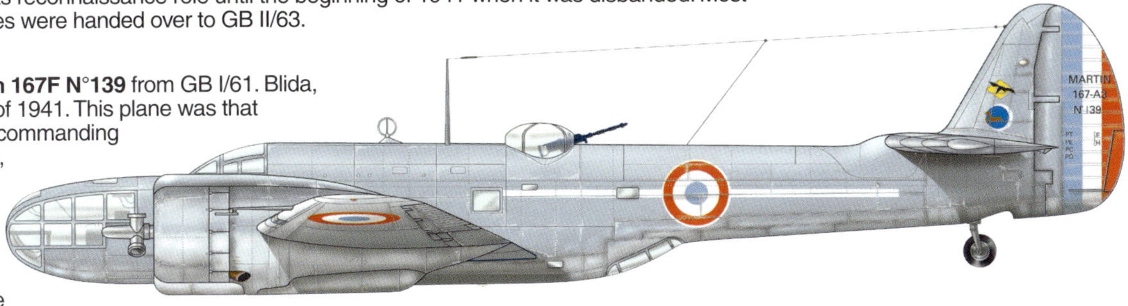

Glenn Martin 167F N°139 from GB I/61. Blida, Algeria, end of 1941. This plane was that of the officer commanding the squadron, witness the two insignia on the tail fin. It has been repainted, the dominant shades being lighter and more suited to its North African theatre of operations. The underside was perhaps aluminium (or light grey-blue) and the upper surfaces dark grey-blue. A circular dated 16 September 1941 made this practice official and stated that the basic shade should be dark grey-blue.

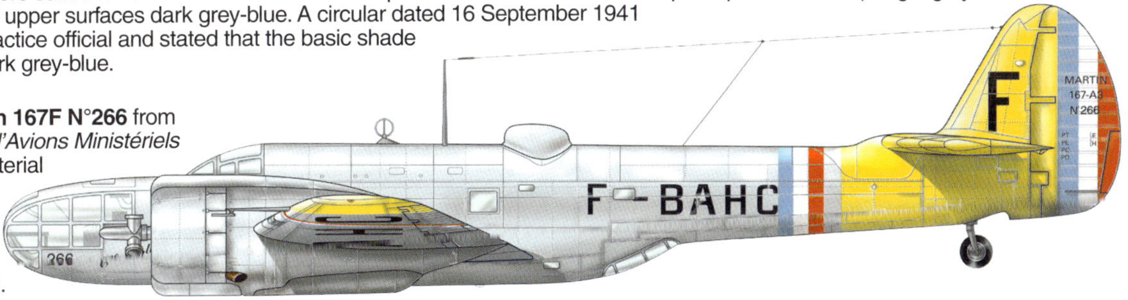

Glenn Martin 167F N°266 from the *Section d'Avions Ministériels* (SAM - Ministerial Aircraft Section). Vichy-Rhue, France, 1942. This plane was christened '*Capitaine Rolland*', painted with special recognition markings and transformed into a passenger transport. It was used on liaison flights to Madagascar via Djibouti, the stop-over in the 'Levant' being replaced by one in Athens in July 1941. The upper gunner's position has been faired over.

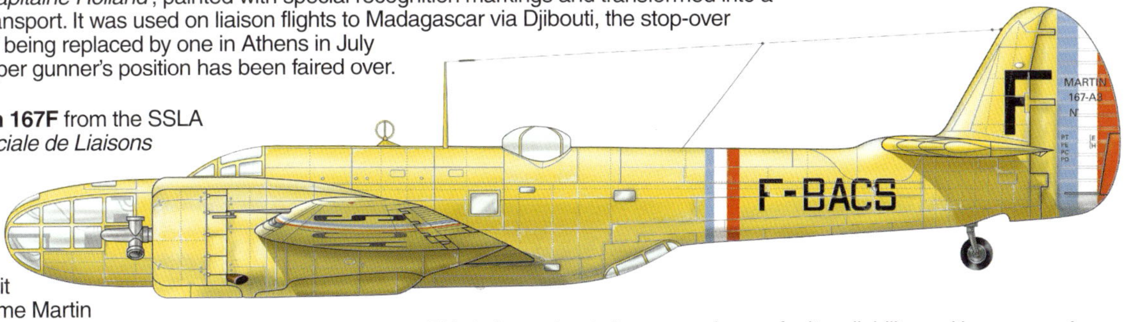

Glenn Martin 167F from the SSLA *(Section Spéciale de Liaisons Aériennes).* Algiers-Maison Blanche, Algeria, 1941. This unit was given some Martin 167Fs specially transformed for rapid passenger transport. This twin-engined plane was chosen for its reliability and long range. In order to avoid any possible mistakes these planes which carried on liaison work between French North Africa and mainland France were painted entirely yellow and wore tricolour stripes around the fuselage and the wings.

Koolhoven FK 58

France was anxious to get hold of the aircraft that its own domestic industry could not produce fast enough. It could not buy any modern machines from Great Britain which was very busy in getting its own air force up to date. After seeking a great deal of help from the United States, France finally turned to the Netherlands. The Fokker Company was not in the market because of its particularly heavy commitment to various programmes (D. XXI, G.1, etc.) so in the end they got in touch with the Koolhoven Vliegtuigen.

The manager, Frederik 'Fritz' Koolhoven, agreed to build a fighter at first designated 'Model 1166', then quickly F.K. 58 (since it was his 58th aircraft project). When the building was well under way, Koolhoven was given a major boost in the person of Erich Schatzi who had just resigned from Fokker where he had created the D. XXI and who now set about contributing in no small way to the completion of the prototype which flew for the first time on 17 July 1938. Less than three months had been needed for the project to be finalised which in itself was a sort of record. The trials with the new machine were conclusive, a second prototype was started. It was slightly different (tail wheel, exhaust pipes, etc.) and series production was planned, even though no official order had yet been placed!

At the beginning of July 1938, the first prototype was taken to France. It was unarmed and bore civilian markings so as not to violate Belgium's neutrality seeing as it had to fly through Belgian air space. It was sent to the CEMA at Villacoublay but there it was incapable of reaching the 312 mph required in the original specifications. It was presented however to the public at the Salon de L'Air several weeks later as the 'fastest fighter in the world' because it had reached 556 mph in a dive during the trials! The FK 58 had a fuselage made of steel tubes covered with canvas at the rear and Duralumin at the front around the engine. The wing was of monobloc construction, made of wood covered with plywood; it went through the airframe - Fritz Koolhoven did not have much faith in the more traditional cantilever construction method.

An official order was placed in January 1939 for fifty FK 58s which were intended as replacements for the Potez 25 TOEs in Indochina and to form two of the three fighter escadrilles for the Far East Armée de l'Air. The power plant chosen was the Gnôme & Rhône 14 N 39 rated at almost 1 100 bhp at take off. The first four production machines were sent to the CRAS in May 1939, bearing Dutch civilian markings but fitted out with Hispano-Suiza 14 AA 10 engines as a matter of necessity. It was in the following July that the first 'real' Gnôme & Rhône-powered FK 58s started to arrive by air. Transferring the aircraft by air made for complications so it was decided to assemble them in France, at Nevers, after they were shipped by boat to Le Havre. The reality was quite different since ten or so planes arrived in France the day before the outbreak of war and the last machines made it to France...by air! In December 1939, 40 FK 58s had been made and 23 were lying unassembled at Nevers, where in the end only one was completed. The Armée de l'Air which refused the four Hispano-Suiza-powered versions (they were perhaps given to the CIC at Etampes) then only had 18 (numbered 1 to 18) at its disposal which it accepted in January-February 1940... and then it did not know what to do with them. Although several examples equipped local defence units (at least three FK 58s served at Caen-Carpiquet alongside the D.510s), two patrols totalling seven Koolhovens were briefly based at Salon de Provence where they carried out 27 operational sorties until 5 June, the date at which all the surviving aircraft were sent to Toulouse and Clermont-Aulnat where they flew twenty or so missions. The last surviving FK 58 fighters were scrapped after the Armistice, ten of them being at the time in the 'Zone Libre', another was captured by the Germans at Villacoublay.

TECHNICAL SPECIFICATIONS

Type
Single-seat fighter (C1)
Power plant
One fourteen cylinder Gnôme & Rhône air-cooled radial engine rated at 1 070 bhp at take-off.
Dimensions
Length: *28 ft 7 in.*
Height: *9 ft 11 in.*
Wingspan: *32 ft 2 1/2 in.*
Wing Area: *186.199 sq ft.*
Unladen Weight: *4 246 lb.*
Max. Weight: *6 050 lb.*
Performance
Max. Speed: *312 mph.*
Cruising Speed: *281 mph.*
Range: *468 miles.*
Max. Ceiling: *35 508 feet.*
Armament
Four wing-mounted 7.5-mm FN/ Browning Model 38 machine guns.

A group of LeO H 257s
from the Aéronavale's Flottille 3B in 1939
(ECPAD 1281)

Lioré & Olivier H 257bis

Although everybody is aware that the French Naval Air Arm (Aéronavale) was often equipped before and during WWII with the same aircraft as those serving in the Air Force (Armée de l'Air), like the D.520, LeO 451 and Glenn Martin 137F, etc., what is less well-known is that the Armée de l'Air itself used seaplanes. When war broke out in September 1939, a bomber squadron was made up of this type of aircraft, using the Lioré & Olivier H-257bis biplane-floatplane.

The story of this machine goes back to the **LeO 20**, one of the great successes of the French aircraft industry in the 1920s-30s. It was a huge biplane capable of carrying its own unladen weight over a long distance at a speed of 125 mph. In answer to the 1928 BN4 programme, Lioré and Olivier presented a fairly similar aircraft, the **LeO 25**, the prototype of which flew for the first time in November of the same year. It was, on the whole, identical to the other machine except for a larger wing, the tail and the power plants (Hispano Suiza V-12s rated at 575 bhp replacing the Gnôme & Rhônes rated at 420 bhp); it carried a four-man crew (pilot, bombardier serving twin machine guns in the front, rear machine gunner and radio operator serving the machine gun covering the lower defensive sector from a retractable bin) and an offensive load which was carried externally under the fuselage and the wings. It performed scarcely better than its forerunner (135 mph in spite of the increased power) and after a series of modifications it was finally abandoned. The six Leo 25 prototypes (three civilian and three military models) were 'navalised' in order to equip the Aéronavale which had been showing an interest in LeO's giant biplanes for some time. These were given a special designation (H-250 to H-255, the letter H standing for '*Hydravion*' - seaplane).

The **H-256**, a new prototype, was made and flew in 1931. It had a longer wingspan and therefore a bigger wing area, new engines (650 bhp Hispano-Suiza 12 Nbrs); its struts were simplified and it was equipped with floats made of Dural and Vedal. This sole prototype gave rise to the **LeO H-257** which flew for the first time in March 1933. The new seaplane had the H-256 airframe but this time had air-cooled engines (Gnôme & Rhône 14 Kbrs rated at 700 bhp) installed in new nacelles, and a new rear fuselage and tail. The wings were no longer swept and the fuselage was enclosed. After the initial trials, a new power plant was installed and the front machine gunner's position was closed and equipped with a cupola. The floats which carried extra fuel, could be replaced by standard landing gear; this configuration reduced the weight by 1 430 lb.

Three official orders were placed between 1936 and 1939 for a total of 58 aircraft, designated **LeO H-257bis** after being given new power plants (Gnôme & Rhône 14Knrs/Kors rated at 870 bhp). They were produced first at Rochefort-sur-Mer then at the SNCASE at Berre. Production was very slow and went on from 1936 to 1939.

The first machines were delivered to the Armée de l'Air which equipped GB II/25 (made up of the Aéronavale's former 4B1 and 4B2 Escadrilles) with them based at Karouba, in French North Africa. Although several machines had been given over to the Navy, 19 Leo H-257bis were still in service in September 1939 and were only replaced by LeO 451s between June and August the following year. This was after it had been envisaged assigning them to two escadrilles of GT I/15 and III/15. When the Free Zone was invaded by the Germans, ten or so of these planes were captured and handed over to the Italians who scrapped them without even using them. The last LeO H-257bis flew in French North Africa with the Allies this time, until 1945.

TECHNICAL SPECIFICATIONS

Type
torpedo bomber twin-float seaplane

Power plants
Two Gnôme & Rhône 14Knrs/Kors Mistral Majors rated at 870 bhp each and driving a Ratier three-blade propeller.

Dimensions
Length: *56 ft 4 in.*
Height: *22 ft 4 in.*
Wingspan: *84 ft 2 in.*
Wing Area: *1 436.860 sq ft.*

Unladen Weight: *11 660 lb.*
Max. Weight: *17 600 lb.*

Performance
Max. Speed: *240 mph at 13 200 feet.*
Max. Ceiling: *19 140 feet.*

Armament
Three 7.5-mm Darne machine guns in the front, ventral and rear positions and one 1 320-lb torpedo or 1 540-lb of bombs carried externally

Lioré & Olivier LeO C.30

After acquiring the licence to build La Cierva autogiros in 1931, Lioré & Olivier which was well-known for its multi-engined planes ordered four **LeO C. 30s** from the British firm A.V. Roe. They were powered by Genet Majors and were shipped to Villacoublay, France where they made various demonstrations. Their excellent handling having been demonstrated, particularly to the public, when a C.30 manoeuvred over the Grand Palais (in central Paris) where the Salon de l'Aviation was being held, an official order was placed on 35 April 1935 with Lioré & Olivier who had opened a special 'rotor wing' department. The order was for 25 C.30 autogiros intended for observation and reconnaissance.

It was only in January 1936 that the first machine came off the production lines, six months late on the initial schedule, a delay which had been caused by problems finalising the airframe. They were originally intended to equip only the Armée de l'Air, but the Marine Nationale (French navy) also showed an interest and four machines from the original batch were handed over to them.

Although the first machines built in France were equipped with 7-cylinder Armstrong-Siddeley engines, the machines built after that were powered by the 9-cylinder Salmson 9 Ne engine. Moreover several modifications of the original design were made by Lioré & Olivier of which the most visible was the presence of little fins under the horizontal tail surfaces to increase the machine's stability. The last example of the initial order was delivered to the CRAS (Centre de Réception des Avions de Série) on 25 July 1938 but further orders were placed in 1937, and on 16 December 1938, 64 C.30s were accepted by the Armée de l'Air which then had 55 autogiros at its disposal of which 33 were ready to fly. A derivative, the **LeO C.301**, flew for the first time in April 1938 followed by an order for five machines.

When war was declared, 52 LeO C.30s were operational, shared out among the CIOAA (*Centre d'Instruction à l'Observation sur Avions Autogires* - training centre for observation with autogiros), some of the GAOs (where fifteen or so were kept in reserve). A further order for 100 machines was placed at the beginning of September 1939 which came to nothing, so that on the eve of the Battle of France, on 10 May 1940, twenty-two C.30s and C.301s were in service with the Armée de l'Air in a number of reconnaissance units, alongside more conventional machines (GAOs 2/520, 1/506, 2/506 and 1/514) and in the CIOAA; moreover, the Aéronavale had nine of these strange little birds in the Escadrille 3S-2 based at Hyères. The operational use of the French autogiros was relatively limited, the aircraft being used for reconnaissance missions before being restricted to close quarter headquarters liaison operations which were made daily more dangerous by the overwhelming German air superiority and because they were unarmed and very slow. When the Armistice was signed, the seven surviving C.30s were captured by the Germans who decided to scrap them since getting them airworthy again was thought to be too costly.

Above.
LeO C.30 n°19. This machine has been painted '*tableau*' green with the engine cowling and rotor head fairing left metal colour. At the time the roundels were only painted on the two elevator surfaces and can therefore not be seen from this angle.
(WingMasters Archives)

Among the derivatives of this autogiro, there was the **LeO C.302**, equipped with a 135 bhp Salmson 9 Nc engine and made in 1938; there was also the **LeO C.34** which was originally intended for tourism, powered by a 350 bhp Salmson 7 Kg engine which was given to the Armée de l'Air, after being exhibited the 1936 Salon de l'Aviation; it finished its career at Villacoublay. There were also two scrapped C.30s which were converted in 1939 by installing a Salmson 9 Ng engine rated at 175 bhp but only one of these machines was ever built... in 1946, and was finally handed over to the Musée de l'Air.

TECHNICAL SPECIFICATIONS

Type
Two-seat reconnaissance and artillery observation post

Power plant
One Armstrong-Siddeley 7-cylinder Genet Major air-cooled radial engine rated at 140 bhp (machines from the beginning of the production series). One nine-cylinder Salmson 9 Ne radial engine.

Dimensions
Length: *19 ft 11 in.*
Height: *11 ft 2 in.*
Tail span: *10 ft 3 in.*
Rotor Diameter: *37 ft 1 1/2 in.*
Unladen Weight: *1 265 lb.*
Max. Weight: *1 892 lb.*

Performance
Max. Speed: *106 mph*
Range: *218 miles*
Max. Ceiling: *13 200 feet.*

Lioré & Olivier LeO 451

Above: **a LeO 451 seen at Toulouse at the time of the Armistice. The Mercier engine cowlings were very close fitting and were specifically designed for this plane. They were equipped with a special system (without flaps) which ensured that the engines were cooled properly.** *(WingMasters Archives)*

The 1934 B5 programme had been started in order to rectify the previous programme's faults, especially the 'multi-role' BCR concept which had been influenced by the theories of the Italian general, Giulio Douhet. In order to do this the future bombers would carry a 2 200-lb to 3 300-lb bomb load, fly at 250 mph at 13 200 feet and, in order to reduce drag, carry the bomb load internally in a bay; the front turret was removed and any others had to be very carefully faired over. At the end of 1935, the crew was reduced from five to four men (the programme changed from B5 to B4), the co-pilot could do the navigator-bombardier's job and rear defence was entrusted to a canon, a retractable ventral bin defending the lower sector. Finally in September 1936, the top speed was increased to 295 mph by an amendment to the initial programme.

Several aircraft manufacturers presented prototype designs (Latécoère 570, Amiot 341, Romano 120 and Bloch 131) whereas within the Lioré et Olivier company, a team led by Jacques Mercier started to design a machine responding to the official programme requirements. This resulted in the **Leo 45 n° 01** prototype which took off for the first time on 16 January 1937. Five weeks later, Lioré et Olivier were nationalised and incorporated into the SNCASE *(Société Nationale de Construction Aéronautique du Sud-Est)*. The machine was powered by two Hispano-Suiza AA 6 and 7 (on the right) installed in standard engine cowlings and driving variable pitch propellers of the same make, that could be controlled in flight. Exchanging the engines for Gnôme & Rhône 14Ps rated at 1 200 bhp was possible without substantial structural modifications and the top speed calculated for this configuration rose to 320 mph. The first trials revealed a certain lateral instability and a tendency to yaw especially on take-off, defects which were rectified by changing the tail fins. Other trials, official this time and carried out at the CEMA from September 1937 onwards revealed that the oil radiators set into the wing leading edges were not effective enough and they were repositioned in the more classic place in the engine nacelles. In spite of these drawbacks, the threshold speed of 300 mph was reached during these trials.

But the prototype's major drawback was without doubt the power plant: it was fragile (they had to be changed every eighteen flying hours) and had a rather defective cooling system, faults which were the cause of a belly-landing in a field near Saclay on 6 December 1937 after the engines had both cut out together. It was made airworthy again very quickly, equipped with Mercier engine cowlings (named after the Chief-Engineer at the SNCASE) which ensured a cleaner aerodynamic shape together with better cooling. With a slight dihedral angle on the outer wing sections, the LeO 45 n° 01 resumed its trials again at the CEMA and an official order was placed in November 1937 for 20 machines to be delivered in May the following year. On 29 August 1938, it was decided to abandon the Hispano-Suiza engines which were too complex and taking too long to finalise with the risk of delaying the aircraft's entry into service, in favour of Gnôme-Rhône 14Ns.

Thus equipped, the plane became the **LeO 451 n° 01** and started its series of official trials again at the CEMA on 21 October 1938. At the same time, the official order was increased, for one hundred extra machines to be produced in the SNCASE factories at Clichy and Levallois, with final assembly at Villacoublay. The order was confirmed in February 1939 even though the production lines had already started; delivery of the first production series machine (LeO 451 n° 01) was however considerably delayed since it only took to the air on 24 March 1939 after being exhibited at the Salon de l'Air in November 1938. The second production series machine (LeO 451 n° 02), powered by GR 14 N 39s driving propellers rotating in the same direction - a feature which was abandoned in the end - flew a month later.

Despite being slightly less powerful, the LeO 451 was capable of the same performances as its predecessor. Its fuselage was made entirely of metal and was of monocoque construction comprising sixty frames fitted with stringers covered with a light alloy outer skin. In order to reduce the master cross-section and at the same time keep the internal gangway giving access to the rear part of the plane, the main bomb bay was reduced and additional bays set into the wing sections between the fuselage and the engines. Each could contain one 220-lb or 440-lb bomb. The tail which had a pronounced dihedral (13°) ended with two fins and was bolted on to the top of the fuselage. The production series machines were delivered, depending on availability, with either GR 14 N 38/39 or 48/49 engines driving propellers by the same manufacturer. But when there were difficulties in supplying the propellers, it was decided to use smaller

LeO 451s outside the factory. The planes have only their nationality markings and have not yet been armed.
(IWM)

diameter Ratier propellerswhich affected its performances, the top speed dropping from 312 mph to 300 mph.

An experimental flight was set up in the spring of 1939 when two new official orders for 100 then 480 machines were placed, first deliveries to the Armée de l'Air starting in August 1939. In September 1939, although 749 machines had been officially ordered (a figure which included five high-altitude LeO 457 bombers, ten LeO 458s equipped with American engines and twelve bombers bought by Greece), only ten machines out of the 22 produced had been delivered to the Armée de l'Air.

The exact numbers of LeOs from the '45' series ordered is difficult to establish and this figure is only of relative importance because only a small number of the planes originally planned (about 450) in fact came off the production lines before hostilities ceased. At the time of the Armistice, more than 1 700 planes had been ordered and several production sites had been opened (at Marignane - its first plane flying in April 1940; at the SNCAO factory at Bougenais. An underground production site had even been planned at Cravant near Auxerre, but this never had the time to operate). Producing the bomber called for more than thirty main sub-contractors among which were Breguet and the SIPA.

The first machines available were assigned to the *31^e Escadre de Bombardement* (Bomber Group) whose GB I/31 started the war with five LeO 451s (alongside eight Bloch 200s) based at Connantre. The initial plan was for ten units (GBs I and II/11, I and II/12, I and II/23, I and II/31 and I and II/31) to be converted. In theory, the two groupes making up each escadre were to receive 27 LeO 451s and conversion and re-equipping was planned to end in February 1940. For this, all the units concerned, except those equipped with Amiot 143s and Farman 221/222s were withdrawn from the front line at the end of 1939 to be converted, being attached to the GIABSE (*Groupement d'Instruction d'Aviation et de Bombardement du Sud-Est* - South-East Bomber Flying Training Group) specially created at Avignon in December 1939. Deliveries to the units suffered from delays because of the indispensable changes which had to be made to the production series machines (special assembly lines were even set up to prevent production from slowing down). The new machine turned out to be difficult, the LeO 451 being very sensitive on the controls and relatively delicate to land. This situation was improved slightly thanks to a training «tour» made to the escadrilles by Jacques Lecarme, chief test pilot at the SNCASE, who had been called up into the Armée de l'Air with the rank of Capitaine.

On 10 March 1940 out of the 106 LeO 451s which had been officially accepted, only 31 were really available. This figure rose to 222 by 10 May, only 94 aircraft (of which 54 were actually available) being in their units. Together with the Breguet 693 in the GBAs, they were the most modern bombers that France had at its disposal. GBs I and II/2 making up the *Groupe de Bombardement N°6*, carried out the first war mission (respectively six and four machines) over Tongres/Maastricht. This squadron was reinforced on the following 15 May by machines from GB II/31, units which went into action on 15 and 16 May before being withdrawn from the front line on the 21st, then sent back into action on the 28th. GBs I and II/1 then went into action (25 May), reinforced in June by GB II/23 from the *Groupement N°7* (GB I/23 did not have its LeOs as planned and flew its antiquated MB 200s...).

On 1 June, *Groupement N°6* which regrouped GB I/12, II/12, I/31 and II/31 went from ZOAN to ZOAE and was transferred to night-time missions because of the losses it suffered during the first weeks of the fighting. It was on this occasion, on 4 June, that the BMW Munich factories were attacked, an operation which was not particularly successful since only two aircraft managed to identify the target because of the bad weather. One plane was shot down. Finally, because of the disastrous military situation, the groups started day-time missions again then, after the 14 June, the LeO-equipped GBs started getting orders to rally French North Africa, the objective being to prevent as many machines as possible from falling into enemy hands and eventually to envisage the possibility of carrying on the fight from the other side of the Mediterranean.

In spite of everything, in the Alpes ZOA, GBs II/11 and I/23 (the latter having finally received its LeOs at the beginning of June) were joined by GB I/25 which had arrived from Tunisia and been converted in France. These squadrons took part in the 13 and 14 June attacks against Italy, which had just entered the war. Three days later *Groupements N°7* (GB I and II/23) and n°11 (GB I and II/11) went over to French North Africa. Only eleven LeO-equipped groups stayed behind in France, including *Groupement N°6* which was used against Italy up to the Armistice, its last mission being on 24 June. It is worth noting that one of the last missions of the war was carried out by four LeOs from GB II/11 which bombed the Sicilian town of Palermo on 23 June after taking off from their new base in French North Africa.

When hostilities ceased, out of the 452 machines built, 373 were delivered and about 130 were lost during the six weeks of the Battle of France. This high attrition rate is perhaps explained by the difference between the plane's intended role and the tasks it was actually given, the LeOs being used like the Breguet 693s, in isolated low-level attacks (whereas the twin-engined bomber was

LeO 451 based at Oran-La Sénia in 1940. The camouflage is very worn down in places, one of the characteristics of this model. The stepped tail fins show that is one of the earlier models. *(SHAA 01877)*

most powerful at 16 500 feet...) which was very dangerous because it exposed the crews needlessly to enemy fire and to the Flak which was often very heavy and deadly. Moreover although theoretically very effective (LeO n°183 from GB I/11 shot down two Me 110s during the same sortie), the rear defensive 20-mm canon was in reality rather ineffective because it was very delicate to reload in flight, and as its magazine held only 60 rounds... However when the LeO 451 was used in the role it was intended for, it was a particularly effective plane. For example, on 16 May, aircraft from GBs I and II/12 and I/31, escorted by fighters, successfully attacked a Wehrmacht division in the Moncornet region.

Among the thirteen bombing groupes which were disbanded according to the terms of the Armistice agreement, three were equipped with LeO 451s (GBs II/12, II/31 and II/11) and a hundred or so aircraft of this type managed to get across the Mediterranean before the fighting ceased. In all, seven squadrons in the new Vichy air force were equipped with LeO 451s, two based in the so-called 'Free Zone' at Istres (GB I/12 and I/31 from *Groupement de Bombardement N°6* which had recovered planes from the disbanded GBs II/12 and II/31) and five based in North Africa: GB I/25 and II/25 (the latter exchanged its LeO 257bis seaplanes for LeO 451s in August 1940) based at Tunis el Aouina; GB I/11 at Oran with GB I/19 and II/61 flying DB-7s (*Groupement N°7*); and finally GBs I/23 and II/23 which formed *Groupement N°11*

LeO 451s from the GB I/12 in flight over Southern France during 1942. *(SHAA B82298)*

at Meknès with I/32 and II/32 which also flew DB-7s.Generally, life in these units under the colours of the Armistice air force was rather peaceful, almost monotonous to repeat the expression of an officer in one of the groups, except in certain cases. Some of the planes in GB I/11 for example attacked the British fleet after the attack on the French fleet at Mers-el-Kebir. Some weeks later, on 23 and 24 September 1940 following Operation 'Menace' against the port of Dakar, a detachment of bombers from GB I/11, I/23, II/23 and II/25 was sent on a reprisal raid to bomb Gibraltar, without achieving anything particularly successful.

What was more important was that planes from GB I/2 and I/31 (sent from France), and the whole of GB I/25 (sent from Tunisia) took part in the Syrian Campaign, in June-July 1941. On this occasion, the LeO 451s, which had already been through several misadventures during their trip (out of the 49 planes which set out only 45 reached their destination), suffered heavy losses (30 aircraft lost, of which five were shot down during the 855 sorties which were carried out) most of them destroyed on the ground by the enemy.

In 1941 several modifications were made to the bombers which were still in service. For example, all the tail fins were enlarged and replaced by fins which were identical to that tried out in March 1940 on n°51; the rear defensive armament was improved by installing twin machine guns in addition to the original canon.

The Vichy government wanted to standardise its ageing fleet of combat aircraft (D. 520 fighters, MB 175 twin engined light reconnaissance bombers and LeO 451 bombers) and manato resume

TECHNICAL SPECIFICATIONS

Type:
Four-seat bomber (B4)

Power plants:
Two Gnôme & Rhône 14 N 38/39 rated at 1 030 bhp each at 15 840 feet (14 N 48/49 rated at 1 060 bhp each) and driving 10 ft 6 in three-blade Ratier propellers.

Dimensions
Length: *56 ft 9 in.*
Height: *17 ft 4 in.*
Wingspan: *74 ft 4 in.*
Wing Area: *731.884 sq ft.*
Unladen Weight: *16 566 lb.*
Max. Weight: *25 080 lb.*

Performance
Max. Speed: *309 mph at 15 840 ft.*
Cruising Speed: *225 mph*
Range: *1 812 miles (1 437 miles with 1 100 lb of bombs)*
Service Ceiling: *29 700 feet.*

Armament
One 20-mm Hispano-Suiza HS 404 canon with 120 rounds (later with twin 7.5 machine guns). One fixed 7.5-mm MAC 34 machine gun in the nose and another in a retractable bin with 100 rounds. 3 080 lb of bombs (of which 2 200 lb in the main fuselage bomb bay).

LeO 451 from GB I/11 in flight at the end of 1941. The tail and engine cowlings have been given the regulation red and yellow stripes. *(SHAA S792987)*

mass-production of the twin-engined bomber at the Ambérieux factory but at the price of very unfavourable conditions (the *Etat français* - the 'French State' - had to supply twice as many aircraft to the Germans as it supplied to its own air force). 225 new planes were ordered, the first of which, bearing the n° 453 flew in October 1941. When the assembly lines started up again only 201 LeO 451s were available; this figure rose to 270 in October 1942, a further 109 machines being supplied the following month, just before the Vichy air force was disbanded. The modernisation of the bomber units had been able to start thanks to production being resumed again, the plans allowing for 18 squadrons to be converted by the summer of 1943. For example, GR I/22 started to exchange its GM 167Fs for LeO 451s in May 1942 thus becoming GB I/22 and it was the last unit to be entirely equipped with the type; the same thing happened to GB I and II/38. GT III/15 was also in the process of being re-equipped when the Germans invaded the 'Unoccupied Zone'.

After Operation 'Torch' (the Allied landings in French North Africa), the planes in the squadrons which had survived, changed sides but their career was relatively brief as they were now a little weary and spare parts were getting increasingly harder to get hold of. In mainland France, the machines captured by the Germans after the invasion of the Unoccupied Zone were handed over to the Luftwaffe and the Italian Regia Aeronautica for training, some of them being converted into transport planes (**LeO 451T**).

Because of its importance in the French arsenal, it was very quickly decided to bring out several versions of the basic LeO 45, versions which were never produced in great numbers - with a few exceptions - mainly because of the dramatic events which occurred when it was being put into service. The **LeO 456** was a version destined for the Aéronavale later designated **LeO 451M** (M for Marine (Navy). 68 examples of this variant which was barely different from the standard bomber were ordered, the first of them being delivered at the end of May 1940. The **LeO 452** was to have been equipped with Hispano-Suiza 14 engines although the only prototype was not built in that form as it had GR 14 N-38 engines. The designation **LeO 453** was reserved originally for the planes equipped with Pratt and Witney R-1830 Twin Wasp radials. No example of this version was ever built, but the designation was used again after the war for the planes re-fitted by the Atelier de l'Air d'Alger (Aircraft workshop in Algiers) with P & W 1830-67s. The prototype of the **LeO 454** was almost finished when the Armistice was signed. Its main feature was that it was to be fitted with Bristol Hercules II engines which used two types of engine cowling, Bristol or Mercier. The **LeO 455** was powered by Gnôme-Rhône 14 R 0/1. Its prototype flew for the first time on 3 December 1939 and was used for a number of armament trials before the Armistice; the trials resumed again in 1942. Finally several of the 451s were converted to 455 status, some of them being used by the Institut Géographique National (**LeO 451 Ph**) or were used as flying test beds.

Although the majority of the LeO 45 variants were distinguished by their power plants, the pressurised **LeO 457** high-altitude bomber was more original because it kept the wings, tail and undercarriage of the original machine but had a new fuselage. The crew was housed in a single cabin situated forward of the wings. The engines were to be Hispano-Suizas and defensive armament comprised six MAC machine guns firing to the rear. One of the particular features of this version was its radiators which retracted into the lower part of the fuselage. Five examples of this model were ordered officially in 1939 and the prototype, unarmed and temporarily fitted with GR 14 N engines to speed up the pressurised cabin trials, was 80% complete when the Germans arrived and shelved the project definitively. The **LeO 458** suffered the same fate. It was powered by Wright Cyclones rated at 1 600 bhp at take-off. Construction of the prototype of this promising version (almost 600 were on order, as well as a specific version for the Aéronavale, the **LeO 458M**) started in the spring of 1940. Finally the **LeO 459** was a liaison version (passenger transport) proposed after the Armistice. It was unarmed and equipped extra tanks in the bomb bay and was renamed **LeO 451 LC** (Long-Courrier). A slightly different version was designed to carry 14 to 21 passengers; its wing bomb bays were converted into luggage storage space with the designation **LeO 451C** (Commercial), this designation being finally taken up for the dozen or so machines which were quickly transformed into transport planes and delivered to Air France.

LeO 451 N° 234 ('6') and N° 3021 (' white 5') from GB I/12 over France, end of 1941. The plane in the foreground has kept the GB II/12 insignia on the tail (GBs I/12 and II/12 merged in August 1940). *(SHAA)*

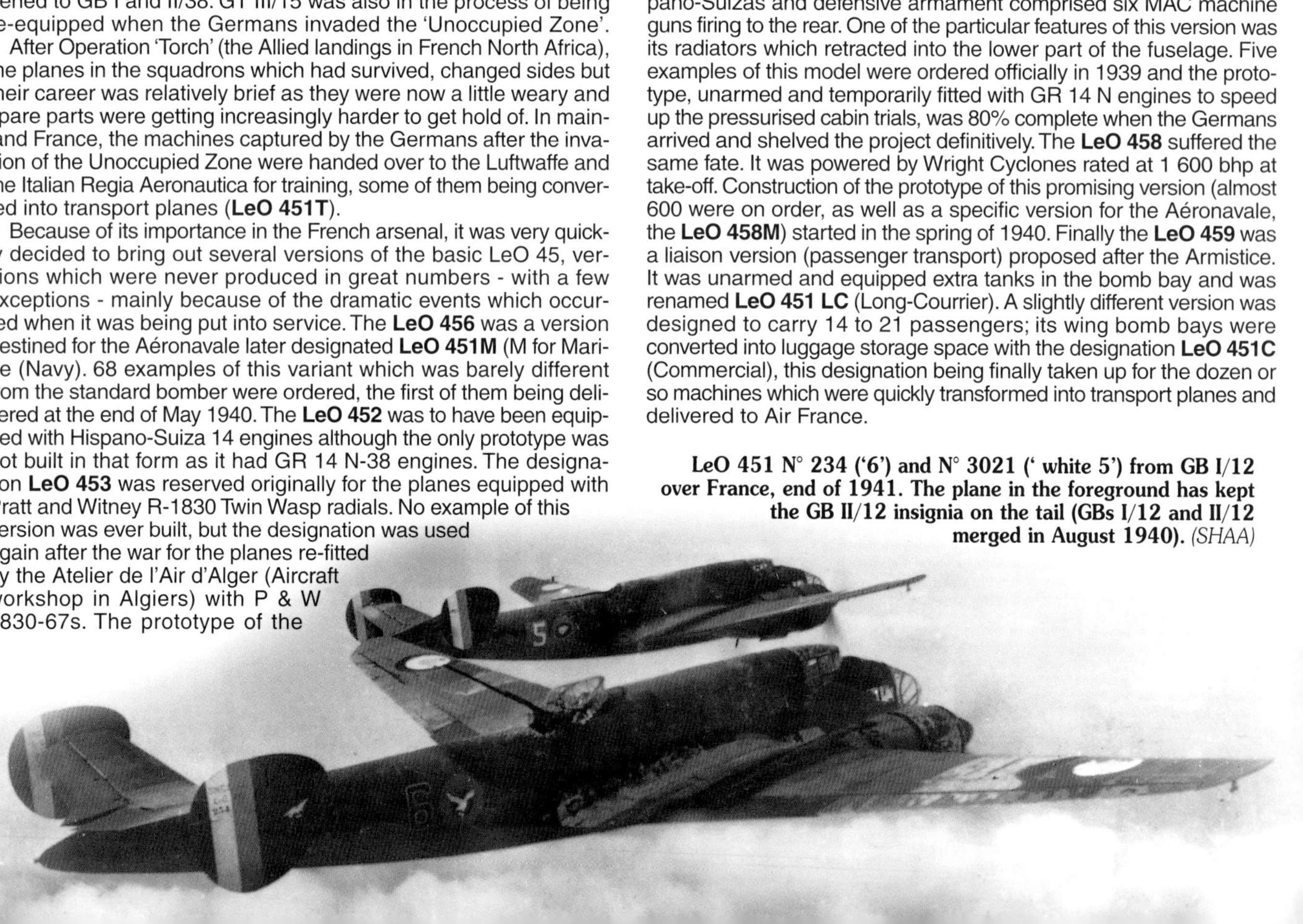

Lioré & Olivier LeO 451

Lioré et Olivier
LeO 451 N°65
from GB I/1. Etampes, France, beginning of June 1940. This aircraft had a particularly long career since it was only withdrawn from service in 1950 after being assigned to GB II/31 based at Istres after the Armistice. Put into service urgently, most of the LeOs used during the Battle of France did not have a unit insignia, let alone an individual number.

Lioré et Olivier LeO 451 N°378 from the 1re *Escadrille* of GB I/11. Oran-la Sénie, Algeria, summer 1941. After the Armistice, GB I/11 together with GB I/9 and II/61 equipped with Douglas DB-7s formed the *Groupement de Bombardement N°3* and took part in the attack on the British fleet after Mers-el-Kebir. This machine, which was delivered in October, was lost by accident on 25 April 1943 after it had gone over to the Allies.

Lioré et Olivier LeO 451 from the 2e *Escadrille* of GB I/11. Oran-la Sénia, Algeria, beginning of 1941. After having carried out its missions on 25 May 1940, the G*roupement N°11* which comprised GB I/11 and II/11 went over to French North Africa on 17 June and stayed there until the Vichy air force was disbanded, in November 1942.

Lioré et Olivier
LeO 451 N°331 from the 1re *Escadrille* of GB I/1. Oran-la Sénia, Algeria, November 1942.
This group took part briefly in operations during the Allied landings in North Africa, one of its bombers being shot down by anti-aircraft fire on 9 November 1942.

Lioré & Olivier LeO 451

Lioré et
Olivier LeO 451
from the *2e Escadrille* of GB I/11. Oran-la Sénia, Algeria, 1942. This plane's colour scheme conforms to the February 1942 directive which banned the use of the roundel and the white fuselage stripe and imposed the addition of tricolour stripes (12 inches wide and at an angle of 30° to the horizontal axis of the machine) on the upper and lower wing surfaces.

Lioré et Olivier LeO 451 n°429 from the *4e Escadrille* of GB II/11. Tunis-El Aouina, Tunisia, August 1940. While getting ready before being sent to Dakar, GB II/11 was stationed in Marrakech before in the end being disbanded on 19 August 1940 and its mission assigned to GB I/63. The «red paper bird» painted on the front of the fuselage was common practice on machines in the *11e Escadre*. The plane's serial number was sometimes added to the insignia, repeated in smaller lettering.

Lioré et Olivier
LeO 451 from the *1re Escadrille*
of GB I/12. Istres, France, March 1941. This group was extensively reorganised in August 1940, its first *Escadrille* taking up the traditions (and especially the insignia, that of BR 134) of the *4e Escadrille* of GB II/12 after it was disbanded. Based in mainland France, the only action this plane ever saw was the Syrian campaign in the summer of 1941. Out of the 49 LeO 451s sent to the 'Levant', all squadrons included, 30 machines were lost during the operations, four during the trip out there and five others shot down in aerial combat.

Lioré et Olivier LeO 451 N°3021 from the *2e Escadrille* of GB I/12. Istres, France, end of 1940. This squadron merged with GB II/12 in August 1940 and was one of the two LeO units based in France after the Armistice. On this plane, the unit insignia has replaced the fuselage roundel. Sent to Syria, this bomber never reached its destination because it was lost in an accident in Italy during the transfer.

Lioré & Olivier LeO 451

Lioré et Olivier LeO 451 n°43 from the 3e *Escadrille* of GB II/12. Persan-Beaumont, France, 13 May 1940. This plane flown by *Capitaine* Knipping, was damaged during the second French attack on the Albert Canal bridges. It was repaired and then transferred to French North Africa. In 1941 the stepped fins characteristic of the early machines were systematically replaced by ones without the «step» on the leading edge.

Lioré et Olivier LeO 451
from the 2e *Escadrille* of GB I/22. Rabat-Salé, Morocco, summer 1942. It was after May 1942 that GR I/22 exchanged its Martin 167Fs for LeO 451s to become GB I/22 in the process but keeping its former insignia. This conversion had been made possible because production of the bomber was resumed in France, the bomber becoming the standard type in the Vichy air force. Almost all the machines in the group were destroyed on the ground during Operation 'Torch', in November 1942.

Lioré et Olivier LeO 451 from the 3e *Escadrille* of GB II/23. Meknès, Moroco, 1941. Planes from this squadron took part in the two reprisal operations against Gibraltar in July and September 1940 where the results were rather more symbolic than particularly effective.

Lioré et Olivier LeO 451 n°424 from the 1re *Escadrille* of GB I/25. Tunis-El Aouina, Tunisia, end of 1941. The paint on the LeO 451s wore off very quickly and many planes were repainted during their careers or even left sometimes in their natural metal colour, either partly or wholly, camouflage no longer being necessary in units which were often far from the combat zones.

Lioré & Olivier LeO 451

Lioré et Olivier LeO 451 from the Headquarters of GB I/31. Quseyr, Syria, June 1941. This plane which bears the special Syrian campaign markings (tail, forward fuselage and engines painted yellow) to avoid possible confusion with Allied planes also belonged to the unit's commanding officer, shown by its individual number ('13') and the particular emblem which comprised the insignia of the two Escadrilles making up the Groupe.

Lioré et Olivier LeO 451 n°142 from the *1re Escadrille* of GB I/31. Istres, France, 1941. This plane was part of the detachment from the squadron which was sent to Syria in June 1941 and which was destroyed on the airfield at Abu-Darme near Aleppo when this was bombed on the following 2 July. Unlike the *2e Escadrille* of this *Groupe*, the *1re* kept its own traditions after GB II/31 was disbanded.

Lioré et Olivier LeO 451 n°388 from the *2e Escadrille* of GB I/31. Istres, France, September 1940. The *2e Escadrille* of I/31 adopted the traditions (and the insignia of BR 226, of course) of the 3e in GB II/31 when it was disbanded in August 1940.

Lioré et Olivier LeO 451 from the *2e Escadrille* of GB I/31. Istres, France, end of 1940. As can be seen, the markings imposed by the Armistice commissions (white stripe, white border of the fuselage roundel) were not immediately painted on all planes. This one has exchanged its fuselage roundel for a large-size insignia, a common practice on the LeOs. GBs I/31 and II/31 merged on 25 August 1940 to form GB I/31 only.

Above:
MS 406 N° 640 from the 4^e^ *Escadrille* of GC II/2 during the Battle of France; this squadron was disbanded in August 1940. The plane no longer has any insignia and its individual number is smaller than the roundel on the fuselage. The disk on the tail is more than likely red. *(SHAA)*

Morane Saulnier MS 406

When WWII broke out in September 1939, the Morane-Saulnier was the most numerous fighter in the Armée de l'Air which is not really surprising since with the Potez 63.11, it was the only aircraft at the end of the thirties whose production reached over a thousand planes. Its origins went back to the C1 (single-seat fighter) programme, intended as a replacement for the Dewoitine D.371/500 and Loire 46 already in service. The programme was submitted to the manufacturers in July 1934 and one of the main requirements was that the aircraft fly at 250 mph at levelling out altitude. The armament was that of the 1932 programme's specification and was made up of various combinations (one canon and two machine guns, two cannon only or simply four machine guns). This programme comprised two categories which had to satisfy the same demands, with only a few special exceptions.

The first category ('C1 *léger*'-light) corresponded to machines which could be made of wood and powered by Renault, Salmson or Potez 400- to 500 bhp engines. The second category included fighters whose mass did not exceed 2.5 tonnes and whose Hispano-Suiza or Gnôme & Rhône engines had to provide 800 to 1 000 bhp.

The programme was changed twice in 1934 before being definitively fixed in November of the following year. The top speed demanded was increased to 280 mph with the necessary range and the service ceiling was fixed at 37 950 feet. Among the different projects that were presented (e.g. Dewoitine 513, Loire 250, Nieuport 161, Romano 130) only two were ever mass-produced: the MB 150 (cf. Vol. I) and the **MS 405**, put forward by the Morane-Saulnier design team. The new type's prototype (n°01), powered by a Hispano-Suiza 12 Ygrs, took off from Villacoublay for the first time on 8 August 1935 with the famous pilot, Michel Détroyat at the controls and temporarily fitted with fixed undercarriage. The trials lasted well into the middle of the next year, the plane revealing very good flying qualities (it reached 306 mph and even 462 mph ...in a dive) and the CEMA only required the maker to carry out minor modifications to the wings and the propeller.

The State officially bought it in November and the prototype continued various trials and even took part in the Brussels Salon de l'Aéronautique in June 1937. Détroyat having brought the machine back from the show in record time (268 mph on average), several countries expressed an interest in the fighter, among which China, Switzerland and Turkey.

A second prototype powered by a different engine (HS 12 Ycrs) driving a propeller with a bigger diameter (which meant increasing the length of the undercarriage) was in turn put into service from the beginning of 1937. The two machines being considered as 'good planes' at the end of their official trials in July 1937 (they were both lost a short time later, N°01 in December flown by a Lithuanian pilot, the other N°02 on 29 July), an official order was placed in 1936 for 15 pre-series machines, to be used to define the future production series fighter as well as the different export versions, and confirmed the following March. This pre-series was produced between February and June 1938, several technical solutions being tried out on different machines. For example, N° 12 was given a

MS 406 N° 732 (Matr. L-761) from the 1re *Escadrille* of the GC I/2 in 1940.
(A. Leblanc Coll.)

Above:
A line-up of MS 405s from the experimental flight set up at Rheims in 1938. The planes have khaki camouflage with grey-blue on the underside. *(Coll. J. Delmas)*

different wing, with a reduced number of reinforcements thanks to the use of Plymax - Duralumin-covered plywood. Then it was fitted with an HS 12 Y 45 engine with a fixed radiator and re-designated **MS 411** (first flight on 24 January 1939).

N° 13 was fitted with a fixed radiator and bays set in the wings for holding flares (modifications which were asked for by Switzerland), and became the **MS 408**, Switzerland buying the licence to build this version in July 1938. As for n°14, it was transformed into the **MS 407Lp** (for Lance-Parachute - parachute launcher) with its fuselage fuel tank replaced by a compartment which could take a parachutist. Three machines were modified like this in 1938-1939 before being re-converted back to classic MS 406s. Finally n°4 powered by a Hispano-Suiza 12Y 31 rated at 860 bhp driving a 10-foot Chauvière propeller in a way became the prototype for the **MS 406** (n°01) and made its first flight on 20 May 1938, the first machine from the production series doing likewise on the following 1 June.

Although the first official order for 50 MS 405s (N°16 to 65) later changed to MS 406s, was placed in April 1937, the mass-production of the MS 406 was decided upon very quickly because of the deplorable condition in which the French fighter strength found itself at the time and also because of the international situation which was worsening daily.

TECHNICAL SPECIFICATIONS

Type
Single-seat fighter (C1)

Power plant
One Hispano-Suiza 12 Y 31 rated at 860 bhp and driving a 10-foot diameter Chauvière 351M three-blade propeller.

Dimensions
Length: 26 ft 11 in.
Height: 10 ft 9 in.
Wingspan: 35 ft in.
Wing Area: 170.593 sq ft.
Unladen Weight: 4 169 lb.
Max. Weight: 5 588 lb.

Performance
Max. Speed: 303 mph at 16 500 ft.
Range: 687 miles
Service Ceiling: 31 020 feet.

Armament
One 60-round 20-mm Hispano-Suiza HS 9 cannon firing through the propeller spinner and two 7.5-mm MAC 34 wing-mounted machine guns.

Above:
MS 406 N° 805 assigned to GC I/7 being unloaded from the seaplane supply ship 'Commandant Teste' in Beirut harbour in February 1940. *(SHAA)*

Below: **MS 405 from the experimental flight seen at the CEMA in 1938. The plane in the foreground is Kostia Rozanoff's, recognisable by the dog 'Dick' painted below the canopy.** *(WingMasters Archives)*

MS 406 from EC 2/596 based at Dong Hoi (Indochina). The insignia on the tail shows a Phoenix, the emblem of the Empress of Annan, the unit's godmother. *(SHAA)*

Production was shared out among various nationalised companies (fuselage by SNCAO at St-Nazaire, wings by SNCAC at Billancourt (near Paris) and the tail by SNCAM in Toulouse) and the final assembly was done by the factory at Bouguenais near Nantes. A second assembly unit was opened later at Puteaux (near Paris) for N°971 onwards. Although production took some time to take off, it gathered speed in 1939, so much so that in the end, although the MS 406 was a complex machine to build (it took 16 000 hours to make, the double of a Bf 109 or even a D. 520), 600 machines came off the assembly lines in the eighteen months following the prototype's first flight.

Various official contracts followed quickly especially for export (China and Lithuania for instance) so that the 1 000-machine figure was reached in 1938. The delivery programme could not be maintained however (572 machines were accepted in September 1939 instead of the 955 planned) in particular because of the delay in receiving deliveries of Hispano-Suiza engines, to the extent where obtaining engines from foreign builders like the Czech Skoda (100 engines were delivered) or even the Russian Klimov, was considered. At the time of the Armistice in June 1940, the *Centre de Réception des Avions de Série* (CRAS) had accepted slightly more than 1 000 MS 406s of which 932 were built in 1939 alone.

An experimental escadrille equipped with MS 405s was made up thanks to the staff of GC II/7 in May 1938 and defined what modifications had to be made to the fighter: improving the landing gear, installing a roll bar inside the cockpit, etc. The *6e Escadre de Chasse* (Fighter Group) was the first to be converted to the new machine, followed by the 2e, 3e and 7e so that when the call-up was announced at the end of August 1939, twelve *Groupes de Chasse* (fighter squadrons) were equipped with MS 406s (GCs I, II and III/12; I and II/6; I, II and III/3; II and III/7 as well as GC I/6 and I/7 based at Blida, Algeria.

When war broke out, the Morane served in seven of the thirteen GCs sent to the front line and carried out almost half of the Phoney War fighter operations, obtaining 32 confirmed kills (out of a total of 81), for 13 machines lost. As soon as it took part in combat, the defects in a fighter which had taken its time entering service were revealed: inadequate protection for the pilot (an armoured plate was installed behind his back as a matter of urgency); defective weapons and radio system; fragile tail skid, which if it broke could cause damage to the tail; badly fitting engine cowling panels which flew off in a tight bend or in dives; and, above all, an under-powered engine which was fragile and wore out quickly, badly affecting its performances... Despite its ease of handling, behaviour and powerful canon, the MS 406 only showed itself to be really effective in the hands of really good pilots who had to make up for its faults with their own skill and talent.

On 10 May 1940 when the Blitzkrieg started, the MS 406 was still in service in spite of the previsions. Even better, it still equipped ten of the twenty-two GCs operating in mainland France with more than 250 aircraft in service, a figure which does not take into account the aircraft used by the training schools, the ELDs and other units of the DAT as well as the machines present in the depots and workshops.

This time the enemy had more Bf 109s - the more recent versions - whereas his bombers, which the Morane could barely catch up with and 'finish off', were anyway often very effectively covered by the cross-fire from guns in other aircraft flying in the same tight formation. This was why all the kills obtained during the Battle of France were obtained at a high price: almost half of all French pilots killed during this period were killed aboard MS 406s. This rate increased when the fighter was used to no great avail in the ground attack role against columns of enemy tanks (GC II/2, I/6 and II/7, particularly on 8 June).

Technical inspection on a MS 406. Note the characteristic angle adopted by the wheels. *(ECPA)*

The MS 406's retractable radiator was replaced by a fixed one on its successor, the MS 410, and its machine guns were fed by belts and no longer from drums. *(DR)*

At the time of the Armistice, about 300 machines had been lost, a figure which takes into account the machines which were abandoned during the dramatic retreat south when units wereobliged to change airfields once a day if not more often... Finally, the MS 406, with its thousand aircraft produced figure (the 1 098 figure, including prototypes and pre-series machines, is generally accepted), was the plane which suffered the highest attrition rate (387, of which about 150 were lost in combat or shot down by German flak). These are figures worth comparing with its score of kills which totalled more than 190 confirmed and 90 probable, some pilots even becoming aces piloting the machine, like Capitaine Williame, from GC I/2 who shot down three Me 109s in a few minutes.

The units flying MS 406s stationed in mainland France were quickly disbanded or re-equipped with D. 520s, the older fighters being handed over to training centres or stored away after having been serviced. A number of them were then used by the Germans who handed some of them over to their allies (Croatia, etc.).

The MS 406's career continued in more exotic climes like the Far-East where it took part in the war against Thailand and especially in Syria in June-July 1941 during which it was opposed to Hurricanes and P-40s among others from the Commonwealth forces. 31 machines were engaged resulting in four kills of which two were confirmed but at the cost of four pilots killed.

At the beginning of 1940 when the MS 406 and the Curtiss H-75 were taking the brunt of the onslaught from the German fighters which were being increasingly replaced by the newer Messerschmitt Bf 109E, it was decided to improve a number of the machines in service (the plane was still being made at the time) and correct some of the defects which came to light in the first months of the conflict. Increasing the armament to four belt-fed (and especially heated, because of the risk of breakdown due to icing up) machine guns was planned, as was replacing the sights and exchanging the retractable ventral radiator for a fixed one, installing ejector-effect exhaust pipes and the standardisation of the Ratier propeller which already equipped some models. The first aircraft thus modified (N° 1028) and called the MS 410 to distinguish it from the other versions, was tried out in April 1940 and turned out to be better than its predecessor with a higher top speed (by 25 mph). Unfortunately this project could not be completed because of the circumstances, the difficulties obtaining supplies and especially because each machine had to be laid up for maintenance for three weeks: during this time they were not replaced by other planes (like the D. 520 which was taking its time entering service). The handful of machines which were being transformed were sent back to their units before the work was even finished. It was only after the Armistice that the MS 410s were used in the units, particularly in GC I/6. Some examples, captured by the Germans were sent to Finland.

MS 406 from the *1re Escadrille* of GC I/7 based in the 'Levant' (French Middle East) in March 1940. *(ECPA)*

Morane Saulnier MS 406

Morane Saulnier MS 406 N° 1031 (Reg. N° L-621) from the *5e Escadrille* of GC III/1, Rouen-Boos, France, April 1940. This machine flown by *Lieutenant* Kazimierz Bursztyn was shot down on 12 May 1940 over Belgium. This unit was mainly made up of Polish pilots who had fled to France after their country had surrendered and so their planes bore the Polish checkerboard in place of the roundels. It is here worn the wrong way round (the red is normally on the top, a mistake which is often made).

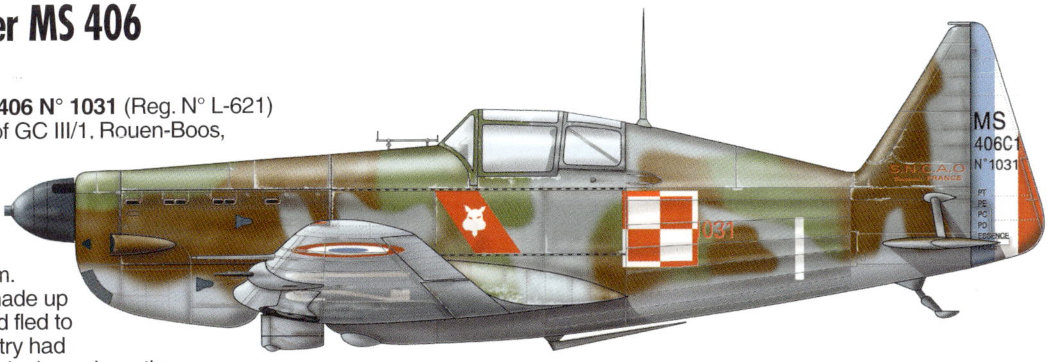

Morane Saulnier MS 406 n°590 (Reg. N° L-529) from the 6e Escadrille of GC III/1. Beauvais-Tillé, France, December 1939. Before the roundel appeared on the fuselage, the planes in this Polish flight bore their insignia (the SPA 93 duck) both on the tail and the fuselage. Afterwards only the tail one was kept.

Morane Saulnier MS 406 N° 410 (Reg. N° L-828) from the *1re Escadrille* of GC I/2. Beauvais-Tillé, France, October 1939. On 4 September 1939, the seven daughters of a farmer (Mr Barbier) living near the airfield where the squadron was stationed became the godmothers of several pilots whose planes bore their first names (here *Jeanine*) above the escadrille insignia, the very famous 'Guynemer's' stork.

Morane Saulnier MS 406 N° 151 (Reg. N° N-471) from the *1re Escadrille* of GC I/2. Beauvais-Tillé, France, September 1940. The Bronzavia exhaust manifold was normally used as a flame damper, especially during night flights.

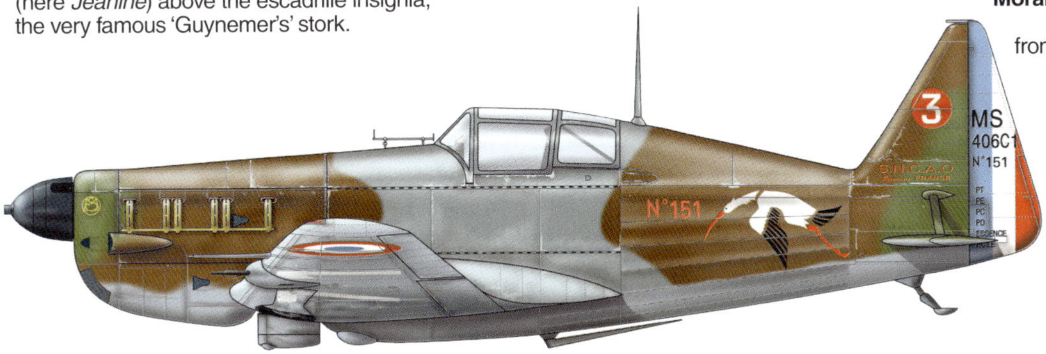

NB.: In order to make the page lay out easier, the ventral antenna is shown folded up when normally it was down in flight.

Morane Saulnier MS 406

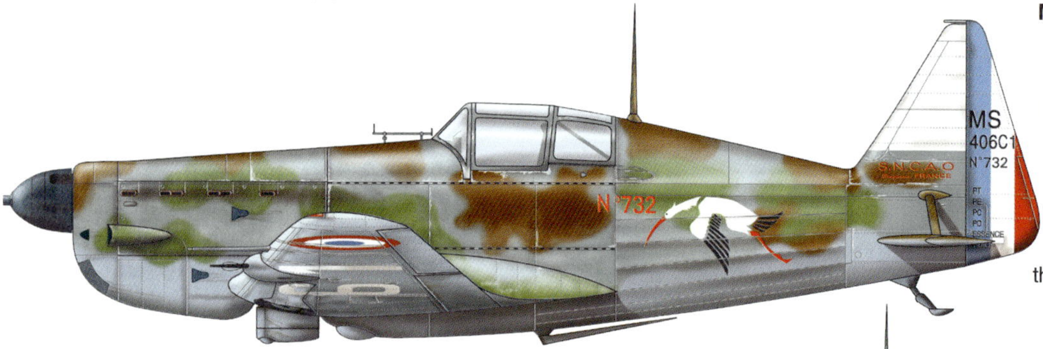

Morane Saulnier MS 406 N°732 (Reg. N° L-761) from the *1re Escadrille* of GC I/2. Beauvais-Tillé, France, end of 1939. This aircraft was assigned beforehand to the Chartres CIC as shown by the white upper part of the tail, white being the specific colour of the French training aircraft.

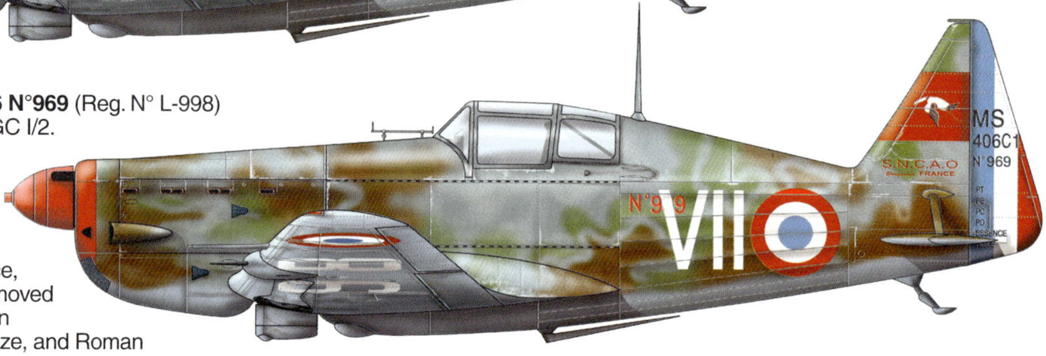

Morane Saulnier MS 406 N°969 (Reg. N° L-998) from the *1re Escadrille* of GC I/2. Nîmes-Courbessac, France, beginning of July 1940. This plane bears the last markings painted on during the last days of the Battle of France, the insignia having been moved to the tail after having been considerably reduced in size, and Roman figures having been used as individual numbers.

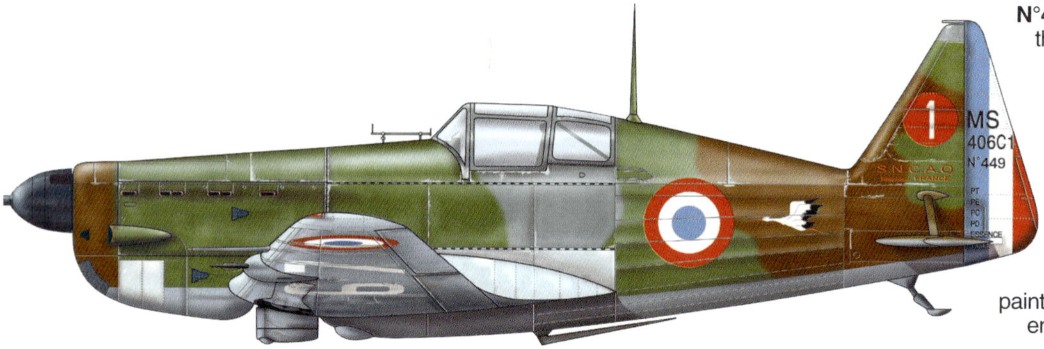

Morane Saulnier MS 406 N°449 (Reg. N° N-867) from the *2e Escadrille* of GC I/2. Toul-Ochey, France, April 1940. Pilot: *Capitaine* Patureau-Mirand. The disk on the tail is red for the second *Escadrille*. The appearance of the fuselage roundel did not mean that the fuselage insignia disappeared immediately; they were painted on the tail towards the end of the Battle of France.

Morane Saulnier MS 406 N°287 (Reg. N° N-699) from the *3e Escadrille* of GC II/2. Soissons-Saconin, France, October 1939. As the individual number painted on the tail (Number '1' inside the black disk of the *3e Escadrille*), this plane belongs to the *Escadrille*'s commanding officer, *Capitaine* Baudoin de Calonnes d'Avesnes.

Morane Saulnier MS 406

Morane Saulnier MS 406 N°923 (Reg. N° L-952) from the 5^e *Escadrille* of GC III/2. Cambrai-Niergnies, France, May 1940. This plane was abandoned on 17 May after being damaged in combat. It was piloted by *Adjudant* Maurice Romey, on of the MS 406 aces with five kills and three probable. GC III/2 converted to Curtiss H-75s at the end of May 1940.

Morane Saulnier MS 406 N°268 (Reg. N° NL680-) from the 1re *Escadrille* of GC I/3. Dijon-Longvic, France, October 1939. This plane which the one belonging to the *Escadrille*'s commanding officer, *Capitaine* Pape, was damaged in 1939 but finished its career in Croatia, to whom it was sold by the Germans in 1943. The planes in the 2^e *Escadrille* wore their red individual numbers on the tail inside a white disk.

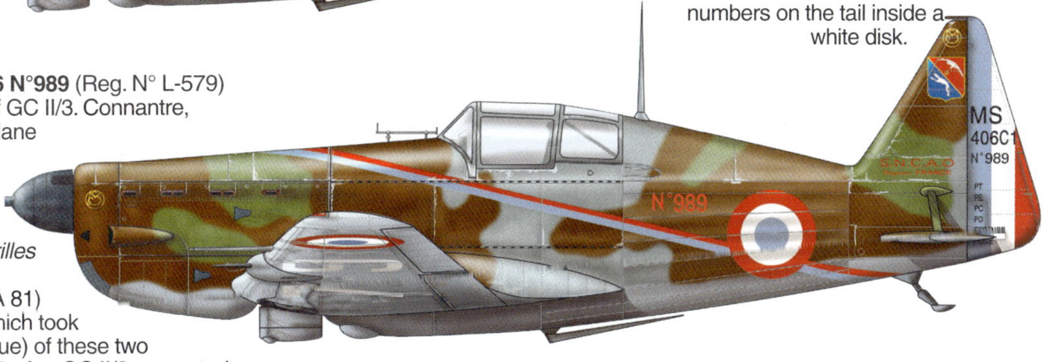

Morane Saulnier MS 406 N°989 (Reg. N° L-579) from the Headquarters of GC II/3. Connantre, France, April 1940. This plane belonged to the commander of the squadron as shown by the blazon uniting the insignia of the two *Escadrilles* (the condor of SPA 37 and the greyhound of SPA 81) and the fuselage stripe which took up the colours (red and blue) of these two units. Transferred to GC I/2 after GC II/3 converted to D. 520s, this MS 406 was shot down on 17 June 1940.

Morane Saulnier MS 406 N°546 (Reg. N° N-964) from the 5^e *Escadrille* of GC III/3. Maubeuge, France, May 1940. The individual number repeated behind the canopy was current practice in the 5^e *Escadrille*, the unit insignia (condor) being apparently only worn by a single plane. This MS 406 was forced to make an emergency landing on 11 May 1940 after being damaged in aerial combat.

Morane Saulnier MS 406

Morane Saulnier MS 406 N°46 (Reg. N° N-357) from the 2e *Escadrille* of GC I/6. Oran-la Sénia, Algeria, June 1939. GC I/6 went over to French North Africa in March 1939 in order to form the French North Africa '*Escadre de Marche*' with GC I/7. It returned to mainland France in the middle of the following December. When the roundel was added to the fuselage, the individual number was put on the fuselage and the *Escadrille* insignia on the tail, but here it was smaller.

Morane Saulnier MS 406 N°413 (Reg. N° N-831) from the 5e *Escadrille* of GC III/6. Wez-Thuizy, France, April 1940. This plane, normally attributed to *Sous-Lieutenant* Martin, was lost during a Heinkel attack in the Dijon region on 10 May 1940. The pilot, *Sergent-Chef* Goujon was able to bale out. Although French planes sometimes had a nickname in 1939-1940, personal insignia like this rather original one on the fuselage, were particularly rare in the Armée de l'Air of the time.

Morane Saulnier MS 406 N°803 (Reg. N° L-832) from the 6e *Escadrille* of GC III/6. Coulommiers, France, May 1940. This plane, nicknamed '*le Dahu*' (an imaginary animal gullible people are supposed to chase), was destroyed in aerial combat on 21 May 1940 in the Cambrai region, its pilot, Sergent Arnould de Gerviliers being killed at its controls. GC III/6, the fourth squadron to be equipped with D.520s was declared to be operational on the new machine at the beginning of June 1940.

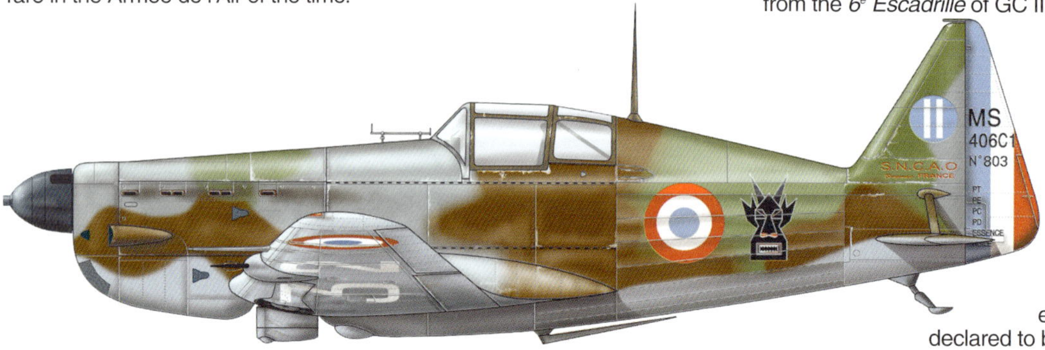

Morane Saulnier MS 406 N°59 (Reg. N° N-370) from the 2e *Escadrille* of GC I/7. Sfax, Algeria, December 1939. when the fuselage roundel appeared on the MS 406s, the unit insignia was moved to the rear of the planes, behind the canopy and greatly reduced in size. When the *Escadre de Marche d'AFN* was disbanded in October 1939, GCs I/6 and I/7 were re-formed.

Morane Saulnier MS 406

Morane Saulnier MS 406 N°899 (Reg. N° L-928) from GC I/7. Estabel, the Lebanon, April 1940. When the planes arrived in the Lebanon, in February 1940, the unit insignia was not immediately painted on the fuselage, only the roundel.

Morane Saulnier MS 406 N°831 (Reg. N° L-860) from the 1re *Escadrille* of GC I/7. Estabel, the Lebanon, May 1940. Like the Martin 167Fs in GB I/39 also based in the Lebanon, the MS 406s in this groups bore Arabic numbers (here '1') as individual numbers. This plane was damaged during the fighting in the Syrian campaign and was the used by the FAFL (*Forces Aériennes Françaises Libres* - Free French Air Force) in GC 'Alsace'.

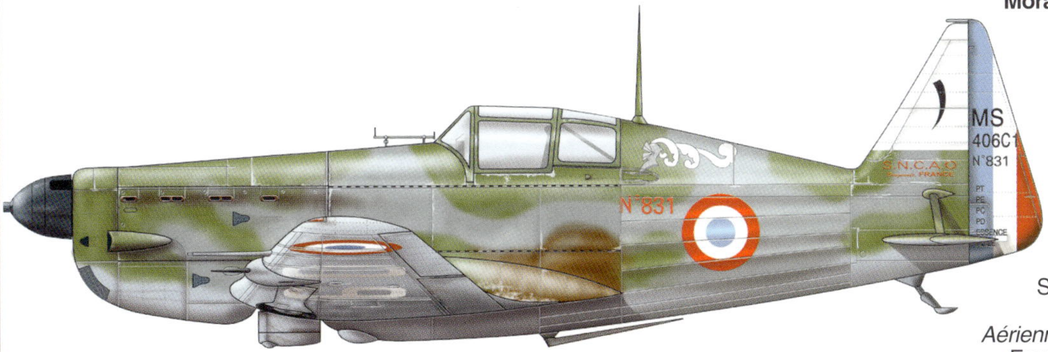

Morane Saulnier MS 406 N°762 (Reg. N° L-791) from the Headquarters of GC I/7. Aleppo-Nerab, Syria, July 1941. This plane, assigned to *Capitaine* Escudier, the executive officer of the squadron, as shown by the double insignia on the tail, was abandoned in Syria then recovered by the FAFL. Like all the planes used in the Lebanon in 1941, the rear was painted yellow, but the horizontal tail surfaces also have red stripes.

Morane Saulnier MS 406 N°204 (Reg. N° N-565) delivered to the 1re Escadrille of GC I/7. Istres, France, June 1941. Stored at Istres, this plane was sent as reinforcement to the Lebanon and was one of the planes which survived the Syrian campaign in the summer of 1941 and returned to mainland France.

Morane Saulnier MS 406

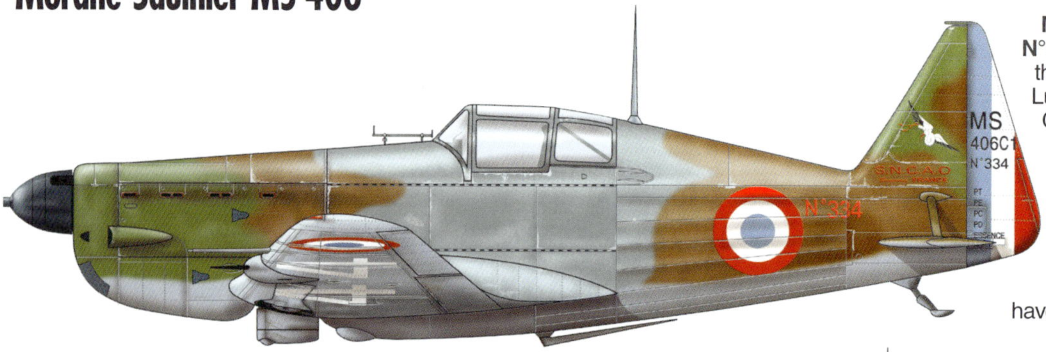

Morane Saulnier MS 406 N°334 (Reg. N° N-744) from the 3e Escadrille of GC II/7. Luxeuil, France, April 1940. GC II/7 which was the first to test the new pre-production series MS 405s in 1938 was re-equipped with D. 520s at the end of May 1940. The planes in the *3e Escadrille* seem no to have had individual numbers.

Morane Saulnier MS 406 N°795 (Reg. N° L-824) from the *6e Escadrille* of GC III/7. Vitry-le-François-Vauclerc, France, beginning of 1940. This plane, piloted by *Adjudant* Jean Bertrand bears the two white vertical stripes showing it was the '*chef de dispositif* (section's leader) plane. The *6e Escadrille*'s insignia showed a 'Fury', taken from one of Michelangelo's frescoes and painted on the fuselage, in spite of the large-sized roundel.

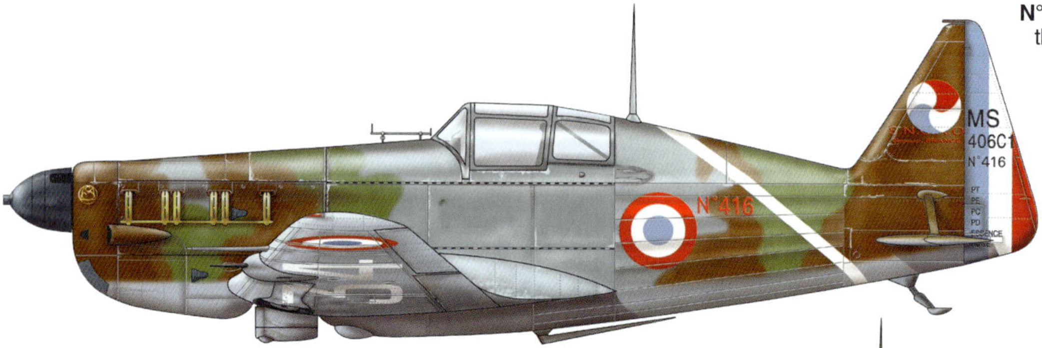

Morane Saulnier MS 406 N°416 (Reg. N° N-834) from the *1re Escadrille* of GC I/9. Oran-la Sénia, Algeria, 1940. This plane bears a stripe on the fuselage indicating that it is a training plane. Its tail bears an unusual insignia (which some sources give as being that of GC I/10, instead of its unit insignia.

Morane Saulnier MS 406 N°828 (Reg. N° L-857) from the *2e Escadrille* of GC I/9. Marseilles-Marignane, France, April 1940. This squadron arrived in mainland France without its planes in April 1940 and stayed there briefly to be re-equipped with MS 406s before leaving for Tunisia at the beginning of the following May. It was one of the three GCs in French North Africa equipped with MS 406s and retained after the Armistice.

Morane Saulnier MS 406

Morane Saulnier MS 406 N°375 (Reg. N° N-793) from the *Centre d'Instruction à la Chasse* (CIC - Fighter Training Centre) . Oran-la Sénia, Algeria, autumn 1940. As the white vertical fuselage stripes show, this plane, no doubt a former GC I/10 plane, was assigned to training duties after the Armistice.

Morane Saulnier MS 406 N°61 (Reg. N° N-372) from GC I/10. Oran-la Sénia, Algeria, September 1940. GC I/10 received the MS 406s which had been assigned beforehand to the 9e *Escadre de Chasse* when this was sent to mainland France. After the Armistice, the MS 406s based in French North Africa were gradually used for training or stored away.

Morane Saulnier MS 406 N°842 (Reg. N° L-871) from EC 565. Ivato-Tananarive, Madagascar, April 1942. This plane, flown by *Lieutenant* Laurant was shot down by three Martlets from No. 881 Squadron during the attack on the island by Commonwealth forces. EC 565 made up of MS 406s from GCs I/9 and I/10 from North Africa made up the *Groupe Aérien Mixte* (GAM) in Madagascar with the Potez 25s and 63.11s from EC 555.

Morane Saulnier MS 406 N°406 (Reg. N° N-725) from EC 2/595. Tong, Tonkin, 1942. This plane's pilot, Capitaine Pouyade escaped in October 1942 and finished the war in the Normandie-Niemen regiment. The French planes in Indochina originally did not have any Armistice markings. The ones seen here were imposed by the Japanese following a tragic incident in January 1942 when French planes were shot down after being mistaken for American fighters.

Morane Saulnier MS 406

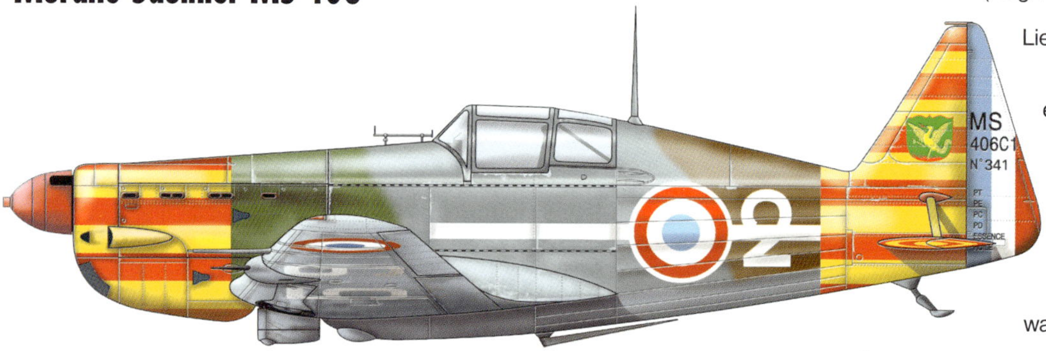

Morane Saulnier MS 406 N°341 (Reg. N° N-759) from EC 2/596. Tong, Tonkin, 1941. Pilot: Lieutenant Hutter. The unit's insignia represents a Phoenix, the personal emblem of Empress Bao-Dai, the unit's godmother and recalls 'Phénix', the French submarine, sunk off Annam at the beginning of the war. ECs 595 and 596 were regrouped at Tong after the base at Nha Trang was ceded to the Japanese in July 1941.

Morane Saulnier MS 406 N°725 (Reg. N° L-754) from the *Ecole de Chasse* (fighter school) at Avord, end of 1939. This plane has white stripes painted on the fuselage and the upper wing surfaces indicating that it was a training aircraft.

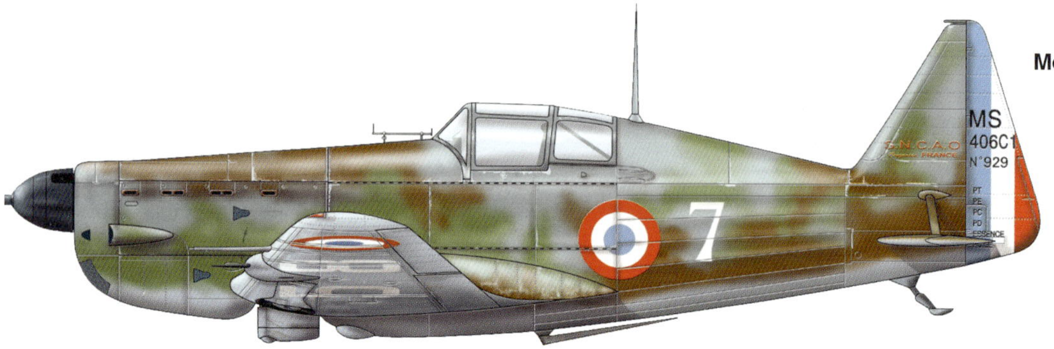

Morane Saulnier MS 406 N°929 (Reg. N° L-958) from the DAT patrol at Montpellier, June 1940. A patrol consisting of four MS 406s and three pilots from the Montpellier CIC was formed early June 1940.

Morane Saulnier MS 406 N°644 (Reg. N° L-673) from the *Escadron d'Entrainement* (training squadron) at Toulouse, end of 1942. MS 406 N° 378 was intended for dive-bombing trials - load of 660 lb being carried and the plane no longer classified as a fighter- was damaged in an accident, so N° 644 was chosen to take its place. The trials very likely did not take place before Vichy's air force was disbanded.

Nieuport Delage NiD 62/622

Above:
A NiD 622 from the 4[e] *Escadrille* of GC II/7 at the beginning of the thirties. These aircraft were quickly replaced in the fighter squadrons but some of them were still on active service in several ERCs at the beginning of the war.
(Coll. J. Delmas)

Although its design dated back a good twenty years, the Nieuport-Delage was brought back into service briefly in second line fighter units which had to be content with this out of date machine or others of the same ilk (mainly the Blériot Spad 510 or the Dewoitine D.510) while waiting for something better.

The story of this machine goes back to the **NiD 42** designed in 1924 as part of the C1 programme launched in 1923. The machine had turned out to be rather unstable and had the rather unfortunate tendency to stall, faults which were rectified by increasing the tail fin surface area. The NiD 42 was thus changed and given the **NiD 52**'s slightly modified wing. Although the new machine behaved rather better, its performances did not necessarily improve, for an all-out weight increase of only 66 lb.

A first official order was placed in January 1928 for a series of modified machines designated **Nieuport-Delage 62 C1**, the very first 'real' NiD 62s -and considered as such from the outset- being delivered the same year. There was a third order in October 1928 for 100 machines. Although the machine, when it left the factory, was almost out-of-date and in spite of the fact that the new C1 fighter programme was launched in 1928, more machines were ordered in 1929 (140 for the Aviation Militaire which had yet to become the Armée de l'Air) and in 1930 (50 machines for the Navy air force soon to become the Aéronavale).

It was soon obvious that the increase in the tail fin surface area which had been made to the NiD 62 was inadequate: the centring was rather too far back and the ailerons were ineffective which caused several accidents. A new wing was therefore installed, the surface of the upper wing being considerably reduced. This new lay-out was tried out on the first NiD 62 N°151 which became the NiD 622, this modification being accompanied by a new engine (Hispano-Suiza 12 Md) which was just as powerful but lighter; it drove a propeller which was now made of metal. There were three orders, for a total of 350 machines which were delivered from the end of the summer of 1931 to the beginning of 1935. A final version, the **NiD 629**, was fitted with a supercharged Hispano 12 Mdsh engine and new landing gear, an order for 50 examples being delivered to the Armée de l'Air in 1935. As soon as it was put into service, the NiD 629 was already out-of-date, compared with its direct competitors, being 31 mph slower than the MS 225 and even 56 mph slower than the D.500!

The front line career of these machines was rather limited since in 1937 all the NiD 62s, 622s and 629s started to be replaced by Dewoitines (371s, 500 and 501s), Blériot-Spad 510s or Loire 46s. It was also in 1937 that the second line air force units were set up, the GARALD (*Groupements Aériens Régionaux d'Aviation Légère de Défense* - Regional Light Aviation Defence Air Groups) which were naturally given these outdated sesquiplanes; 84 of the 233 machines in service in the Armée de l'Air in 1938 were spread out in seven of these units; the others equipped schools or were used in fighter instruction squadrons.

As the GARALDs were replaced by the ERCs from October 1937, the Armée de l'Air had 107 NiD 62/622s (30 operational) at its disposal at the time of the General Mobilisation (end of August 1939), spread out among the ERC 1 and 2/561, based at Rouen (together with D. 501s), ERC 3/561 at Calais (with Spad 510s) and ERC 1/562 at Lyon (which replaced its machines from September 1939 with D. 501s). Moreover, less than forty or so NiDs were in French North Africa in three ERCs (571, 572 and 573). The ERCs were gathered together in the GARC which ended up forming the GCs. ERCs 1 and 2/561 became GC II/10; ERCs 3 and 4 changed into GC III/10 and ERC 1/562 became GC III/9. This reorganisation was followed by the conversion of the mainland units onto MB 151 and 152s, those in French North Africa receiving D.501s (ERC 571 and 573 becoming GC III/4 in June 1940), Blériot-Spad 510 and MS 406s (ERC 572 and 574 becoming GC III/5 on 15 May 1940). Withdrawn from the front line units, the NiD 62/622s were used as trainers or for general duties in non-combattant units in Mainland France or French North Africa, the 17 surviving machines inventoried after the June 1940 Armistice being finally scrapped.

TECHNICAL SPECIFICATIONS

Type
Single-seat sesquiplane fighter
Power plant
One Hispano-Suiza 12b V-12 engine rated at 500 bhp.
Dimensions
Length: *24 ft 9 in.*
Height: *10 ft.*
Wingspan: *39 ft 7 in.*
Wing Area: *332.038 sq ft.*
Unladen Weight: *2 855 lb.*
Max. Weight: *4 026 lb.*
Performance
Max. Speed: *147 mph at 16 500 ft.*
Range: *562 miles.*
Max. Ceiling: *27 060 feet.*
Armament
Two 12.7-mm machine guns in the fuselage.

Potez 25

Above:
A group of Potez 25 TOEs from the 2^e *Escadrille d'Indochine* in the middle of the thirties. This unit became the *Escadrille d'Observation* I/42 in 1939. The extra fuel tanks, a feature of this version, can be clearly seen under their fuselages, painted all-green, except for the engine cowling.
(SHAA 902171)

At the Salon de l'Aviation which was held at the end of 1924, the Potez Company presented a two-seat observation sesquiplane (A2), derived from the Potez 24, with fixed undercarriage, of mixed construction (wood and metal), with fabric covered wings and powered by a double-Vee 12-cylinder Lorraine 12 Eb engine: the Potez 25. The plane was designed in answer to a programme for a two-seat reconnaissance aircraft which could also carry out light bombing missions and be fitted with different power plants (Gnôme & Rhône, Salmson, Renault, Lorraine, Hispano-Suiza, Farman, Clerget, Rolls-Royce, Panhard, etc).

The first prototype, the Potez 25.1, made its maiden flight at the beginning of 1925 and the official trials carried out at the same time were so conclusive that it was decided to launch a pre-series of machines. A second prototype (Potez 25.2 N°1) which was slightly different since it had no rear gun position -replaced by a passenger compartment fitted with a little windshield- was designed almost simultaneously and reflected the manufacturer's desire to offer the model for export which was where in the end it was most successful. At the end of 1925, a hundred or so machines had been ordered, a figure which continued to increase since more than 2 450 Potez 25s of all types were delivered to the Armée de l'Air and to the Aéronavale between 1926 and 1934; 1 500 machines were exported, either made in France or licence-built in other countries like Portugal, Yugoslavia or Romania. At the beginning of the thirties, the Potez 25 had a worldwide reputation, took part in a number of both military and civilian trophies and expeditions and even entered history by being used by the Aéropostale pilots.

After the first unit trials were carried out in 1926, service in the Armée de l'Air started the following year, the planes being delivered to the observation or reconnaissance groups as well as to the training schools. Replacing the Breguet 19 which had been definitively retired from service in 1934, the Potez 25 had a long career, in spite of the arrival of the Breguet 270s and Potez 390s. It was the arrival of the ANF Les Mureaux 110 which tolled the knell for the Potez 25 after 1935 although eight flights were still equipped with the type the following year. It was the CAR, renamed the GAR in 1937 made up mainly of reservists, which received the downgraded Potez 25s. As these GAR became GAOs at the end of 1938, there were no longer any units made up solely of Potez 25s when the war started in September 1939; the machines were dispersed among a number of units where they were used mainly for liaison and training.

Although at the time of the Armistice in June 1940 there were 177 Potez 25s of all versions (and there were many, simply because of all the different power plants it could take) most of these planes had been retired from front-line duty after November 1939 mainly because of their great vulnerability, demonstrated by the type's only combat loss, when a machine from GAO 2/508 was shot down by a Messerschmitt 109 on 27 September 1939.

The most famous version of this, one of the French aircraft industry's greatest successes, was without doubt the Potez 25 TOE (for *Théâtres d'Opérations Extérieurs* - foreign theatres of operations) since this variant represented almost a third of all the Potez 25s which came off the production line. Intended for use in difficult conditions, the machine had been 'tropicalised' with reinforced lubrication and cooling systems for the power plant and an extra fuel tank installed in the fuselage which increased its range quite considerably. Various types of operations could be undertaken by this aircraft, depending on the equipment: day- or night-time observation, bombing (up to 880 lb of bombs could be carried with the fixed armament removed) and 'penetration' (special equipment and increased fuel capacity). The Potez 25 TOEs were found in large numbers in the GAOs in French North Africa (about 160 examples in May 1940) where after the Armistice they were gradually replaced by the 63.11. Others like the ones in Madagascar, Djibouti, French West Africa or East Africa had even longer careers and the machines in the escadrilles in Indochina lasted until 1945.

TECHNICAL SPECIFICATIONS POTEZ 25 (TOE)

Type
Two-seat observation (light bombing) sesquiplane (A2)

Power plant
One 12-cylinder water-cooled Lorraine-Dietrich 12 Eb engine rated at 450 bhp.

Dimensions
Length: *30 ft.*
Height: *11 ft 6 in.*
Wingspan: *46 ft 1 1/2 in (upper wing).*
Wing Area: *495.098 sq ft.*
Unladen Weight: *3 366 lb.*
Max. Weight: *4 730 lb.*

Performance
Max. Speed: *129 (126) mph at 6 600 ft.*
Cruising Speed: *106 (112) mph.*
Range: *475 (468) miles.*
Service Ceiling: *22 110 feet.*

Armament
One fixed forward-firing Vickers 7.7-mm machine gun in the fuselage; twin Darne 7.5-mm machine guns in the rear gun position; one Lewis 7.7-mm machine gun firing downwards through a panel in the fuselage; four 110-lb bombs under the wings and four under the fuselage.

Potez 25

Potez 25 N° 515 abandoned in France in June 1940. As indicated by the red and white checker board lozenge painted on the fuselage, this plane was used for instruction in instrument flying, a hood (here folded back) covering the student's cockpit at the front. The instrument flying emblem was always big so that it was very clear what type of training missions the aircraft was engaged in; they were however simpler than the one shown here.

Potez 25 N° 1485 from EO 1/595. Bach Mai, Tonkin, September 1939. The Potez 25s TOE were equipped with extra radiators and fuel tanks, and bigger wheels in order to take the extra weight. The Potez in Indochina had a particularly long career, some surviving until 1945.

Potez 25 N° 2049 from the *4e Escadrille* of GB II/15. Rheims-Coucy, France, April 1940. On top of the numerous GAOs, the Potez 25s were also used in certain units as headquarters, inspection and liaison aircraft. This aircraft bearing the insignia of BR 132 on its fuselage as well as the white training aircraft stripe, served in the group alongside Farman four-engined bombers.

Potez 25 TOE N°1482 from EO I/41 (*1re Escadrille, Observation*, in the *Groupe Aérien Mixte* - combined air group). Tong, Tonkin, March 1942. This plane bears the markings (white vertical stripe, all the tail painted red) imposed by the Japanese from February 1942.

POTEZ 39

Above:
Potez 39 N°01 seen at Istres at the beginning of the thirties. The machine is not camouflaged except for the flag painted on the rudder. The structure, which was made entirely of metal, the corrugated metal skin of the fuselage and the 'parasol' wings can be clearly seen here.
(Coll. V. Gréciet)

At the end of the twenties, the general staff of the Aéronautique Militaire, which had not yet become the Armée de l'Air, was thinking of replacing the Potez 25s and Breguet 19s which equipped its observation units. So it was within the context of the official 1928 programme that two parallel competitions were launched, one for a two-seat reconnaissance aircraft (R2) and the other for a two-seat observation plane (A2). The specifications for the latter stipulated that the machines be equipped with the Hispano-Suiza 12 Hb V-12 engine and be able to take off in less than 400 feet which meant that the wings had to be particularly large.

The Potez Company decided to take part in the second competition by offering a high wing cantilever monoplane made entirely of metal - which was a first for the manufacturer - and able to carry 4 950 lb and cover 470 miles at a height of 24 750 ft. The machine, designated **Potez 39**, was powered by the engine required in the specification, giving it a speed of 143 mph at 11 500 ft.

The new machine took to the air for the first time in January 1930 and underwent a particularly long and thorough series of trials covering more then 800 flying hours. The plane was chosen by the authorities at the same time as the Breguet 27 (cf. Volume 1, page 56) and one hundred machines were ordered and equipped the reconnaissance groups. The **Potez 37** was presented in the same programme but in the R2 competition. This was a strange machine powered by a Hispano-Suiza 12 Hdr engine rated at 520 bhp whose fuselage ended with a cylindrical cone-shaped boom which supported the tail section, a lay out intended to give the observer an unobstructed field of vision and to avoid blind angles for the twin guns in the rear turret. Since the ANF-Les Mureaux 110 (cf. Vol.1, page 22) was chosen in the end, only three prototypes of this machine were made.

The first Potez 39s which entered service at the beginning of 1934 were assigned to the 32[e], 34[e] and 54[e] Escadres, respectively based at Dijon and le Bourget. According to the original plans, 83 machines were to have been in service by the spring of 1934 but only 76 of them had been delivered by 1 January 1935, the last example being delivered the following May. The Potez 39's career within the *Groupes d'Observation* (GO) was very brief because all of them were converted to newer machines after 1936: GO II/32 was given Mureaux 110s, GO I/34 and II/34 Amiot 143s and GO II/54 Potez 540s.

When war broke out in September 1939, fifty Potez 39s were still in front line units, but also in schools and miscellaneous centres; eleven others were stocked at EAA 301. Seven GAOs (504, 505, 510, 511, 512, 513 and 517) used some of these machines, alongside others, which were often just as outmoded, like the Breguet 270 in particular. It was their very vulnerability which caused *Instruction Générale N° 12* dated 6 October 1939 to be issued ordering the ten GAOs equipped with Potez 39s and Breguet 270s to be withdrawn from the front line sectors.

Assigned to duties away from the battlefield or just stocked away, some of the Potez 39s were nevertheless present in their original GAO units on 10 May 1940, this time alongside their successors from the same firm, the Potez 63.11s, which formed the backbone of the observation units of the time. Completely out of date, the 41 aircraft of this type inventoried in July 1940 after the Armistice, were finally scrapped, putting an end to a lacklustre career.

TECHNICAL SPECIFICATIONS

Type:
Two-seat observation monoplane
Power plant:
One 580-bhp Hispano-Suiza 12 Hb engine.
Dimensions
Length: 33 ft.
Height: 11 ft 3 in.
Wingspan: 52 ft 8 in.
Wing Area: 376.705 sq ft.
Unladen Weight: 3 282 lb.
Max. Weight: 4 950 lb.
Performance
Max. Speed: 150 mph.
Range: 500 miles.
Max. Ceiling: 23 100 feet.
Armament
One synchronised 7.7-mm Darne machine gun under the engine cowling, twin Lewis guns in the rear gunner's position, 264 lb of bombs.

Koolhoven FK 58, NiD 622 & Potez 39

Koolhoven FK 58 n° 11. Salon de Provence, France, beginning of June 1940. This machine is well-known because it was photographed a lot. After being assigned to one of the two patrols at Salon-de-Provence made up of Polish pilots, it was then sent to Aulnat and abandoned in Montpellier at the end of June. Its serial number, first used as its individual number and therefore painted on the fuselage, was then rubbed off as shown on a period photograph. It seems that the FK 58 did not have standard French camouflage; a dark coat (probably brown) was simply painted over the grey-blue upper surfaces.

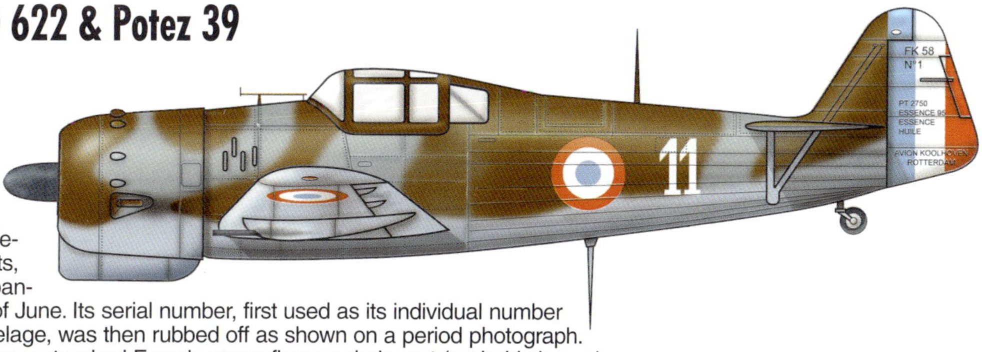

Nieuport-Delage NiD 622 n° 219. The number of this plane, seen in a plane cemetery in June 1940, after the Armistice, is given without any guarantee. It is possible that it had already been used by ERC 3/561, based at Calais-Marck which was also equipped with Dewoitine D. 501s and which finally became GCII/10, after converting to Bloch MB 151/152s.

Potez 39 from GAO 504, Cannes le Bocca, October 1939. This is one of the rare planes of this type to have its upper surfaces camouflaged. Most of the Potez 39s still in service at the time were painted 'tableau' green with perhaps the metal engine cowling painted over to be less visible.

Koolhoven FK 58 n°17. Montpellier, France, June 1940. Most of the twenty or so French FK 58s were mainly used by Polish pilots from the DIAP in Lyons in four patrols based at Salon-de-Provence and Clermont-Aulnat. At the end of its brief career it had its engine cowling repaired (the small bumps were removed and the front opening was widened). The planes in these units which were still available at the end of June 1940 were abandoned at Montpellier and Perpignan when the pilots were evacuated to Great Britain.

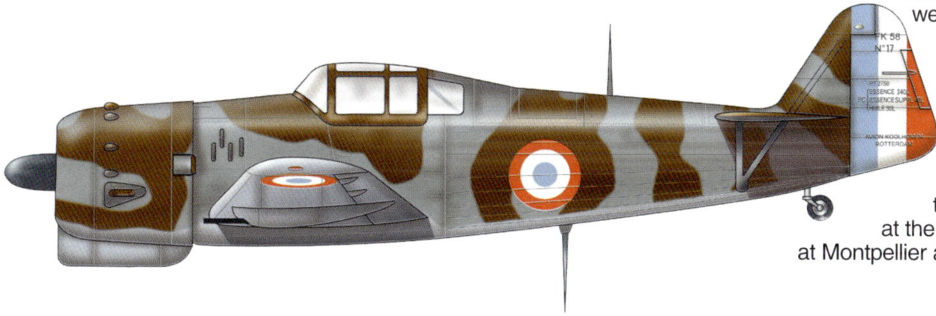

POTEZ 540/542

Above:
This profile shot of Potez 540 N°3 (serial N° X-076) from the 1re *Escadrille* of GR II/55 clearly shows the plane's angular shape and the three gunners' positions intended to give it absolute protection...
(Coll. J. Delmas)

We have already mentioned the Italian General Douhet and his theories which greatly influenced the BCR programme launched in France at the beginning of the thirties. These concepts probably seemed attractive at first because one multi-role machine could be used in place of six different types, but in the end it gave rise to a series of aircraft which were incapable of carrying out one, let alone six, of the missions for which they were designed and which could be summed up in the rather terse expression: good for nothing, bad at everything...

In 1932, the BCR programme was brought up during a private conversation between General Denain, the then Air Minister and Henri Potez, before it was revealed to French aircraft manufacturers in April of the following year. Although some of the manufacturers' interest in producing an aircraft which could carry a 1 100 to 2 200-lb bomb load at 220 mph over 315-440 miles was probably rather mitigated Potez, now forewarned, threw himself into the adventure at his own expense and started a prototype designated **Potez 54 M4** on 25 August 1933. It fitted the official specification and flew for the first time on the following 14 November, finished its official trials on 5 May 1934; deliveries started on 25 November of the same year.

This plane which was made in record time, had a fuselage made of wood and canvas with a rectangular cross-section; the high wing had a metal structure covered with canvas and its outer sections had a dihedral angle; the engines were installed in nacelles into which the landing gear retracted; the nacelles were fitted to stub wings and linked to the main wings by means of struts. It also had gun positions in two cupolas and a retractable bin all armed with a 7.7-mm machine gun, placed in such a way that all the sectors covered by the guns overlapped, without any blind angles, theoretically providing complete protection, as laid out in the 'flying fortress' concept typical of the BCR programme. A bay held part of the bomb load and the rest was fitted on pylons under the stub wings. The Potez was finally chosen after competing with the Bloch 131 (cf. Vol. 1, p. 31), the SAB 80, the Breguet 460 and the Farman 420. At the end of the year, 43 machines were ordered together with a 54.2 prototype, followed the next year by four further orders.

In all 196 examples of this machine came off the production lines in two main versions, the **Potez 540** (146 aircraft) and the **Potez 542** (50 aircraft), most of them being built under licence by Breguet, Latécoère or UCA (Blériot) since Potez did not have enough equipment. In fact the two versions only had different power plants, the 540 having Hispano-Suiza 12 Xbrs air-cooled engines and the 542 having Lorraine Pétrel 12 Hdrs/Hers or Hfrs/Hgrs. The **Potez 541** had Gnôme & Rhône radials and was intended for Romania but was never built whereas 13 examples of the **Potez 543**, which was in fact a variant of the 541, were built for Romania. The last variant, the **Potez 544** was a mail plane equipped with HS 12 Ybrs engines; only one was built.

The first machines to be built were assigned to the GR I/54, I/33 and I/55 and at the beginning of 1937, 187 planes were spread out among 25 escadrilles. This number increased to 244 in 26 escadrilles the following year. At the beginning of the year, and although it had been clearly demonstrated that the aircraft was totally inadequate, the Potez 54 was still to be found in six reconnaissance squadrons of the 33e, 36e and 52e *Escadres*, in four bomber squadrons (25e, 54e and 63e *Escadres*) and in some GAOs (506, 551, 552 and 553) in mainland France, Africa and even Indochina, some of the machines having been modified to serve overseas (Potez 540 and 542 SOE - 3 and 20 machines respectively).

These machines were very rapidly removed from front line service since they were totally outclassed in all the operations for which they were originally intended. The Potez 54s were mainly used to familiarise the Potez 25 crews with 'modern' machines, in this case eventually the Potez 63.11, its designated replacement, alongside other outmoded machines like the Bloch 131 or the MB 200. At the time of the Armistice, 45 machines of both versions were inventoried, 25 others were in Morocco, 40 in Algeria; the Potez 54s belonging to the units based in Indochina were used up to 1941.

TECHNICAL SPECIFICATIONS (POTEZ 540)

Type
Multi-seat bomber, fighter and reconnaissance aircraft

Power plants
Two Hispano-Suiza 12 Xbrs air-cooled V-12 engines rated at 690 bhp at 12 870 feet (542: two Lorraine Petrel liquid-cooled engines rated at 720 bhp at 11 550 feet).

Dimensions
Length: 53 ft 5 in.
Height: 12 ft 10 in.
Wingspan: 72 ft 11 in.
Wing Area: 817.988 sq ft.
Unladen Weight: 8 327 lb.
Max. Weight: 13 156 lb.

Performance
Max. Speed: 187 mph.
Cruising Speed: 150 mph.
Range: 750 miles.
Max. Ceiling: 33 000 feet.

Armament
Three 7.5-mm machine guns mounted in turrets. Ten 110-lb bombs carried internally, four 500-lb bombs under the lower wing stubs.

Potez 542

Potez 542
N° 258
(Matricule Registration number - X-599) from EB 2/42. Tan-Son-Nhut, Indochina, 1940. This machine along with two others on this plate, was equipped with Bronzavia manifolds mounted on the exhaust pipes as flame dampers. All the Potez 540/542 were painted dark-green (the so-called 'tableau' green) all over, except for the engine nacelles which were left a natural metal colour.

Potez 542
N° 112
(Reg. N° X-186) from the *4e Escadrille* of GR II/52. Nancy, France, September 1939. During the Battle of France, this group was only equipped with Potez 63.11s and 637s.

Potez 542
N°120
(Reg. N° X-594) from the *3e Escadrille* of GR II/33. Nancy-Etain, France, end of August 1939. This famous reconnaissance squadron was one of the rare units to have some of the first available MB 174s.

Potez 542
n° 257
(Reg. N° X-598) from EO I/41. Dong Hoi, Indochina, 1943. This unit was disbanded in March 1943. As well as the markings imposed by the Japanese, the plane bears the famous red and yellow 'Vichy stripes' painted on the engine cowling.

Potez 630, 631, 633 & 637

With more than 1 100 aircraft built in five different versions, the Potez was the most numerous French plane produced in the 1935-40 period, in front of the Morane Saulnier 406. Designed to carry out a variety of roles, from fighter to bomber including reconnaissance, it was affected by a defect which plagued it during the whole of its operational career: the lack of power from its engines.

At the end of October 1934, the authorities launched a plan for a 'multi-seat light defence aircraft' which complemented the C1 programme for fighters issued several months earlier. The future plane had to be able to carry out three main missions (fighter command three-seater, two-seat night-fighter, and escort/day-time fighter), its top speed was set at 280 mph at 13 200 feet, altitude to be reached in less than fifteen minutes and its 200 mph cruising speed was to give it an endurance of four hours. In order to be manoeuvrable enough, its total weight was not to exceed three tonnes, a demand which meant one of two power plants: either 450-bhp inline Renault-Salmson engines, or 600-bhp Hispano-Suiza or Gnôme & Rhône radials. Finally armament was to comprise at least two 20-mm cannon and a machine gun for rear defence.

Several firms were interested in the project but they all had to make a prototype at their own expense. In competition with the Hanriot 220 (which became the NC 600), Loire-Nieuport 20, Romano 110 and Breguet 690, Potez presented a machine designed by a team led by the engineers Coroller and Delaruelle, the **Potez 630** N°01. It flew for the first time on 25 April 1936, one year after building had started. It was powered by Hispan-Suiza 14 Hbs engines and had a twin tail made of wood without dihedral in order to test different configurations. Modified following its first flights and powered by new engines (Hispano 14 Ab 02/03 then 10/11 enabling it to reach 287 mph fully laden), the prototype was sent to the CEMA then to Cazeaux for armament trials.

The initial official specification required the use of Gnôme-Rhône engines, so a second prototype, the **Potez 631** N°01, fitted with GR 14 Mars engines which performed less well but were smaller, was also built and flew for the first time almost a year later, on 15 March 1937 before being sent in its turn to the CEMA the following November.

These two versions were identical, except for their power plants, the only difference visible from the outside being on the engine cowlings where the air intakes were on the top on the 630 and

Opposite:
The reconnaissance Potez 633 was very similar to the fighter version, except for the ventral gondola (clearly visible here), the glazing under the tip of the nose and the reduced armament (one machine gun only). *(Coll. S. Bonnaud)*

Above:
Potez 630 N°5 in flight, no doubt at the very beginning of 1938. The plane is not camouflaged and has only the individual number on the fuselage. The air intake on the engine cowling distinguishes this version from the following ones. *(Coll. J. Delmas)*

below on the 631. These were twin-engined monoplanes with a cantilever wing; the fuselage had an oval cross-section and was built in three sections, the central part including the first part of the wing. The wing had a stressed metal skin and fabric covered movable surfaces. The long cockpit canopy protected the two or three crew members, depending on the version and the tail plane had a big dihedral angle and ended with twin tail fins. Armament was that specified in the specifications, even if the two fuselage cannon were not always installed and even sometimes replaced by machine guns.

A first order for ten prototypes which were to be used to define the different versions of the twin-engined plane was placed in May 1937 followed several weeks later by two further official orders. At the time, the Potez Company had been nationalised and incorporated into the SNCAN and it was at the new corporation's various factory sites that production started in May 1937. Final assembly was carried out at Les Mureaux. In all 80 Potez 630 C3s came off the production lines in February 1938.The first production series machine (N°8) was delivered to the Armée de l'Air in the following May and the first 631 on 2 August 1938.

The Potez 630s were delivered to GCN III/1 at Etampes and II/4 at Rheims and to every fighter group in order to be used as 'fighter command aircraft' (the unit commanding officer, installed in the

Potez 631 from the *Escadrille de Chasse de Nuit 3/13* whose insignia has been painted on the nose. The fuselage stripe covers the serial number painted behind the roundel and the individual number has been painted on both surfaces of the tail fin. The propellers have been fitted with their spinners. *(SHAA 792182)*

THE DIFFERENT VERSIONS OF THE POTEZ 63

630: *three-seat fighter (C3) and fighter command. Hispano-Suiza 14 Ab. Two forward-firing 20-mm cannon installed under the belly.*

631: *identical to 630 except for Gnôme-Rhône engines. 215 built.*

632: *two-seat night-fighter (CN2). Not built because converted to two-seat bomber (B2). Hispano engines. One prototype.*

633: *two-seat bomber (B2). Gnôme-Rhône engines. Identical to 631. Reduced offensive weapons and bomb bay. 71 built of which 35 exported.*

635: *two-seat night-fighter (CN2) derived from 631. Radio operator's position removed, two extra cannon. Never built.*

636: *two-seat fighter version (C2) derived from 633 to be built under licence by Avia in Czechoslovakia. Project abandoned.*

637: *observation three-seater (A3). Armament and engines of 633, ventral gondola. 60 ordered.*

639: *assault two-seater (AB3) derived from the 633. Extra armour, bomb launcher underwing, one single 20-mm canon set at 14° for strafing. One prototype.*

63.11: *three-seat observation (A3). Identical to 631 but modified, glazed f ront fuselage and new cockpit canopy. One 7.5-mm machine gun under the fuselage. 723 delivered before the Armistice. A further 120 finished in 1941.*

63.12: *three-seat fighter (C3) made from 631 N°40 fitted with Pratt and Witney. One example converted.*

63,13: *two-seat dive-bomber (Bp2). Dive brakes, two 20-mm cannon. Wing pylons. One machine converted.*

TECHNICAL SPECIFICATIONS (POTEZ 631)

Type
Three-seat twin-engined fighter (C3)

Power plants
Two fourteen-cylinder air-cooled Gnôme-Rhône 14 M4/5 (6/7) radial engines rated at 700 bhp each at take-off and driving Ratier (Gnôme & Rhône) propellers.

Dimensions
Length: *36 ft 6 in.*
Height: *11 ft 11 in.*
Wingspan: *52 ft 10 in.*
Wing Area: *351.950 sq ft.*

Unladen Weight: *5 390 lb.*
Max. Weight: *9 900 lb.*

Performance
Max. Speed: *276 mph at 14 850 ft.*
Range: *762 miles.*
Service Ceiling: *29 700 ft.*

Armament
Two forward-firing 30-mm Hispano-Suiza HS 9 cannon under the fuselage; one 7.5-mm MAC 34 defensive machine gun in the rear cockpit (some machines were equipped with twin 7.5-mm machine guns under each wing).

twin-engined plane co-ordinated the activities of the single-seat fighters by radio), replacing the venerable Bloch MB 200s in this role. But these planes were withdrawn from the front lines at the beginning of the war because of recurring engine trouble, after having become day fighters; they were used thereafter only as training aircraft.

On the other hand, the Potez 631 had a much longer if not truly brilliant career, mainly again because of its inadequate performances. Gradually replacing the Potez 630, 200 631s were built of which 117 were in the front line when war broke out, spread out among GCN I/13 at Etampes (18 machines), GCN II/13 at Rheims (21) and five *Escadres de chasse* - fighter wings -(49 planes assigned to the 1re, 2e, 3e, 7e and 8e ECs), eight others being stationed in Corsica and Djibouti.

Since the units had been organised during the Phoney War, on 10 May 1940, the Potez 631s which were still in service had been shared out between some of the Aéronavale's flottilles, an *Escadrille de Chasse Multiplace* - Multi-seat fighter squadron - (ECM I/16 created in January 1940 with the twin-engined planes assigned to the *Escadres*) and five night-fighter *Escadrilles* which were now autonomous and entrusted with the defence of the Lyons (ECN 5/13) and Paris regions. On 17 May 1940, the latter took part in a day-time attack to halt the advance of the German columns which ended with the loss of two of the aircraft committed. At the time of the Armistice, the Potez 631 units (including the Aéronavale) had made 17 kills in aerial combat, but only ECN 1/13 and 3/13 were maintained, based at Nîmes-Courbessac, the latter being transferred to Gabès, Tunisia in June 1941 in order to reinforce GC II/7 and its D. 520s within *Groupement de Chasse N°24*. At the end of 1942, when the Vichy air force was dismantled, part of the surviving Potezs (six 630s and 64 631s) were handed over to Romania, one of Germany's allies which used them as target tugs or trainers.

The production series two-seat bomber, the Potez 633 B2, was identical to the fighter (same Gnôme & Rhône engines) except for a couple of modifications: the glazing in the nose of the prototypes was reinstalled after having been abandoned in the production aircraft; a single forward-firing machine gun was installed on the front right-hand side of the fuselage; and especially, a bomb bay containing eight 110-lb bombs replaced the navigator's position.

The Armée de l'Air showed an interest in this version in March 1938 (125 were ordered for Plan V) before cancelling its order a few months later, only to replace it partly with 117 Potez 631s. France received several examples intended originally for Greece (13 out of 24 ordered), Romania (19 out of 44) and China (4 out of 9 which

Potez 631 N°162 from ECN 3/13 in flight, used as a two-seater. The tube visible under the fuselage behind the stirrup step is the wire antenna which was deployed in flight. *(SHAA 953824)*

were sent to Indochina), the machines being held back because of the worsening political situation.

The French Potez 633s were used by the assault groups instead of the Breguet 691s/693s they were expecting to be delivered, and served to train the crews in the *54e Escadre* (GBA I and II/54) based at Salon-de-Provence, then those of the 51e and finally in March 1940, GBA II/35. After 10 May, the Potez 633s were used to bring certain units up to strength like GBA II/51 which moved up on 19 May with eight of these aircraft alongside five Breguets. On 20 May, three Potez 633s carried out a single mission in the Arras region and the planes were withdrawn from font line service two days later, six planes having been lost for a variety of reasons before the end of the same month.

The last version of this prolific line, the **Potez 637 A3,** a tactical army cooperation and reconnaissance three-seater first flew in the summer of 1938. The plane used the Potez 631/633 airframe and engines but a partly-glazed ventral gondola was added under the fuselage and fuel capacity was increased and cameras were installed at the rear. Armament was that of the 633 with a MAC machine gun firing forward and another in defence at the rear, a third being added later in the gondola. Finally, a 110 lb bomb could be carried under each wing.

60 machines were ordered in August 1938 before the prototype's first flight which only took place in October. These planes were to replace the Potez 540/542s in four GBs and to make sure that the transition went smoothly while waiting for the Potez 63.11s. First deliveries were made to GRs I/33 and II/33 in May 1939 then GRs I/52 and II/52, GRs I and II/36 receiving their Potez 637s to support the Potez 540s. The first weeks of the war having demonstrated how inadequate the MB 131s (Vol. I, page 31) were for the missions they were given, only the *Escadrilles* equipped with Potez 637s were allowed to cross the enemy lines to carry out daylight reconnaissance missions. During the Phoney War, the Potez 637 was the only modern plane at the disposal of the GRs, with the

Above:
Potez 631 in ECN 3/13 bearing the markings of the Vichy air force on the engine cowlings and the tail. *(SHAA 85074)*

Potez 540/542s, even Potez 25s (in GR II/39 in Syria) being used to bring units up to strength. As the Potez 63.11s were gradually put into service from November onwards, the 637s were progressively withdrawn and assigned to training; this measure was rather unpopular among the crews who considered the plane better to fly and slightly faster than its successor. When the Blitzkrieg began on 10 May 1940, the Potez 637s had more or less disappeared from the front line units; some machines were still to be found in GB II/33, and GB II/54 was the only unit not to have any MB 174s. Having lost all its planes in two weeks, it was given some 63.11s at the end of the month as reinforcements before being transferred to Bordeaux and equipped with MB 175s. Six machines were still on the books in 1st November 1941, but a year later when the Vichy air force was disbanded there were no longer any Potez 637s on the inventory of available planes.

After the Munich crisis, certain machines were camouflaged (dark green) on their upper surfaces like these Potez 630s with a rather spectacular 'serrated' livery. *(SHAA 933498)*

Potez 630 & 631

Potez 630 N°44
(Reg. N°C-556) from the *3e Escadrille* of GC II/1. Buc, France, September 1939. The Potez 630s which were used as fighter command aircraft in the *Escadres* (nine in the 1re EC in September 1939) were very quickly withdrawn after the beginning of the war and used mainly for training. The camouflage here consists only of two colours: khaki (green) on the upper surfaces and grey-blue on the underside. The big escadrille insignia ('death reaping'/ SPA 94) has been painted on the fuselage.

Potez 630 N°45 (Reg. N°C-557) from the *1re Escadrille* of GC I/5. Bordeaux-Mérignac, France, June 1940. This fighter command-plane which kept its very particular khaki camouflage on its upper surfaces was abandoned at Bordeaux at the time of the Armistice. As can be seen here, the official markings (roundels and fuselage stripes) have been successively added, partly covering each other. In September 1939 eight of these machines were used by the *5e Escadre de Chasse*.

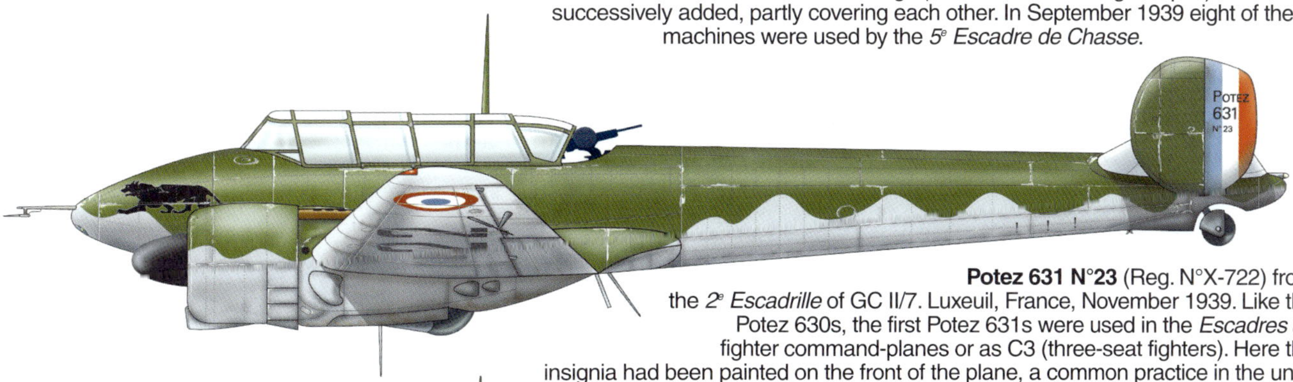

Potez 631 N°23 (Reg. N°X-722) from the *2e Escadrille* of GC II/7. Luxeuil, France, November 1939. Like the Potez 630s, the first Potez 631s were used in the *Escadres* as fighter command-planes or as C3 (three-seat fighters). Here the insignia had been painted on the front of the plane, a common practice in the units equipped with these twin-engined aircraft. The camouflage with a wavy borderline (including on the leading edges) was typical of the machines put into service before the war.

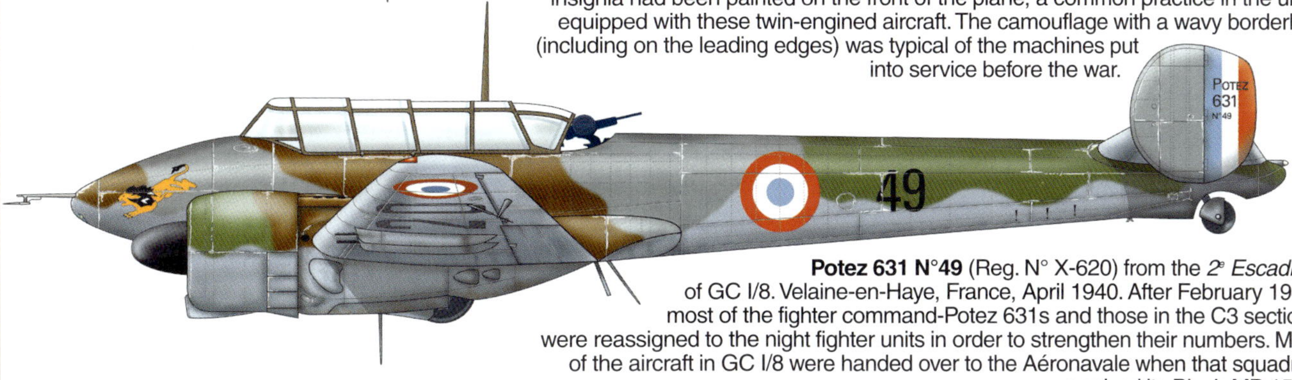

Potez 631 N°49 (Reg. N° X-620) from the *2e Escadrille* of GC I/8. Velaine-en-Haye, France, April 1940. After February 1940, most of the fighter command-Potez 631s and those in the C3 sections were reassigned to the night fighter units in order to strengthen their numbers. Most of the aircraft in GC I/8 were handed over to the Aéronavale when that squadron received its Bloch MB 152s.

Potez 631

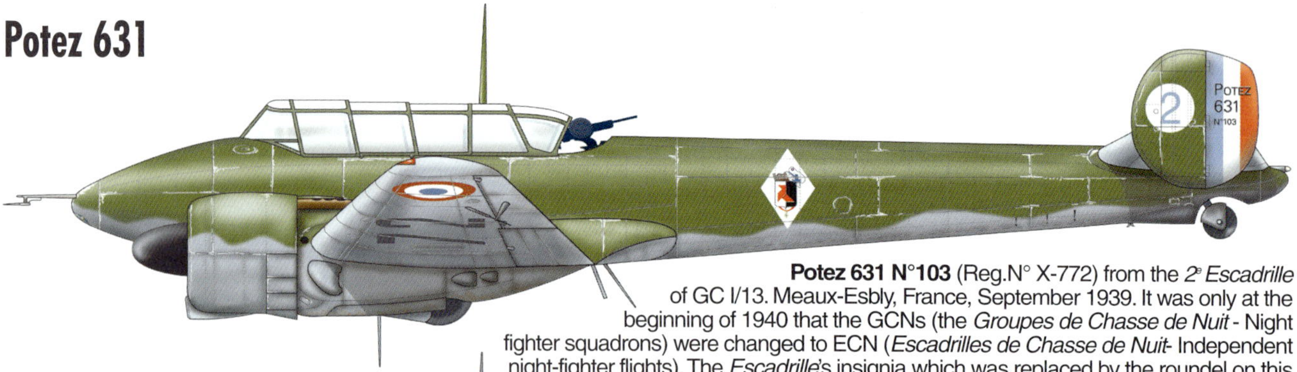

Potez 631 N°103 (Reg.N° X-772) from the *2e Escadrille* of GC I/13. Meaux-Esbly, France, September 1939. It was only at the beginning of 1940 that the GCNs (the *Groupes de Chasse de Nuit* - Night fighter squadrons) were changed to ECN (*Escadrilles de Chasse de Nuit*- Independent night-fighter flights). The *Escadrille*'s insignia which was replaced by the roundel on this part of the fuselage represents a red paper bird over the arms of the city of Metz. On the Potez, the individual numbers were often painted on both sides of each fin.

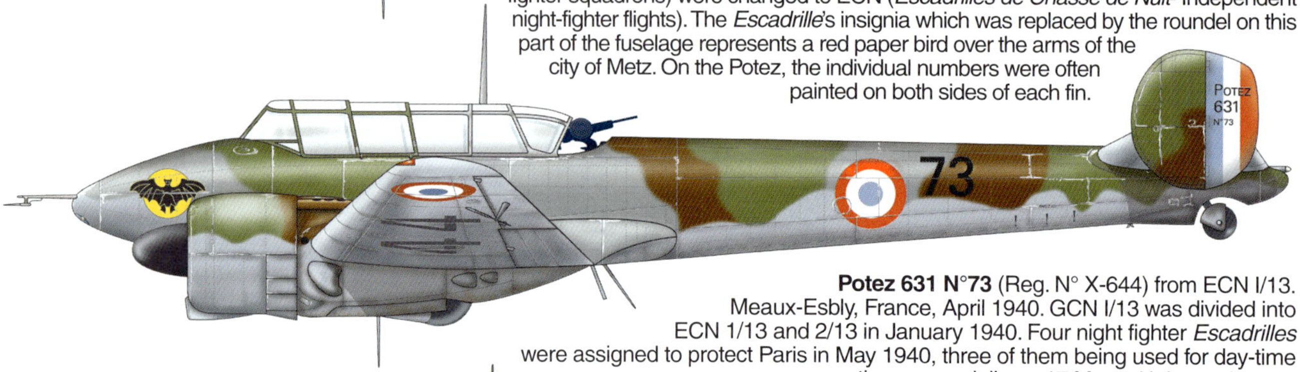

Potez 631 N°73 (Reg. N° X-644) from ECN I/13. Meaux-Esbly, France, April 1940. GCN I/13 was divided into ECN 1/13 and 2/13 in January 1940. Four night fighter *Escadrilles* were assigned to protect Paris in May 1940, three of them being used for day-time operations, especially on 17 May, with heavy losses.

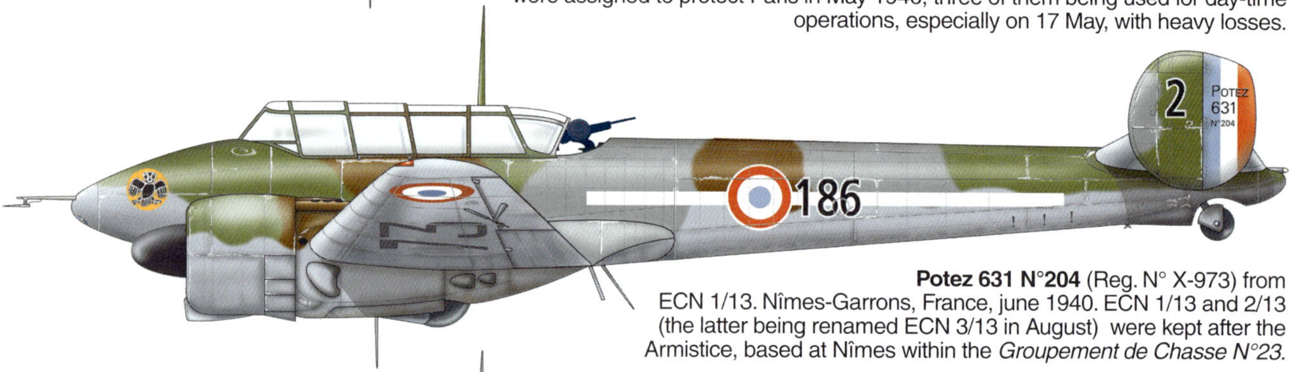

Potez 631 N°204 (Reg. N° X-973) from ECN 1/13. Nîmes-Garrons, France, june 1940. ECN 1/13 and 2/13 (the latter being renamed ECN 3/13 in August) were kept after the Armistice, based at Nîmes within the *Groupement de Chasse N°23.*

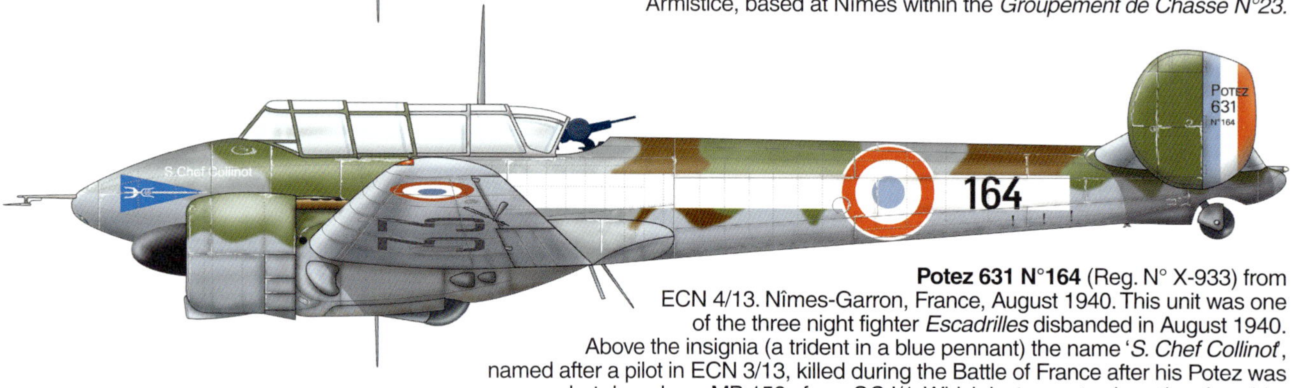

Potez 631 N°164 (Reg. N° X-933) from ECN 4/13. Nîmes-Garron, France, August 1940. This unit was one of the three night fighter *Escadrilles* disbanded in August 1940. Above the insignia (a trident in a blue pennant) the name '*S. Chef Collinot*', named after a pilot in ECN 3/13, killed during the Battle of France after his Potez was shot down by ... MB 152s from GC I/1. Which just goes to show that the white fuselage stripe was not too much !

Potez 631, 633 & 637

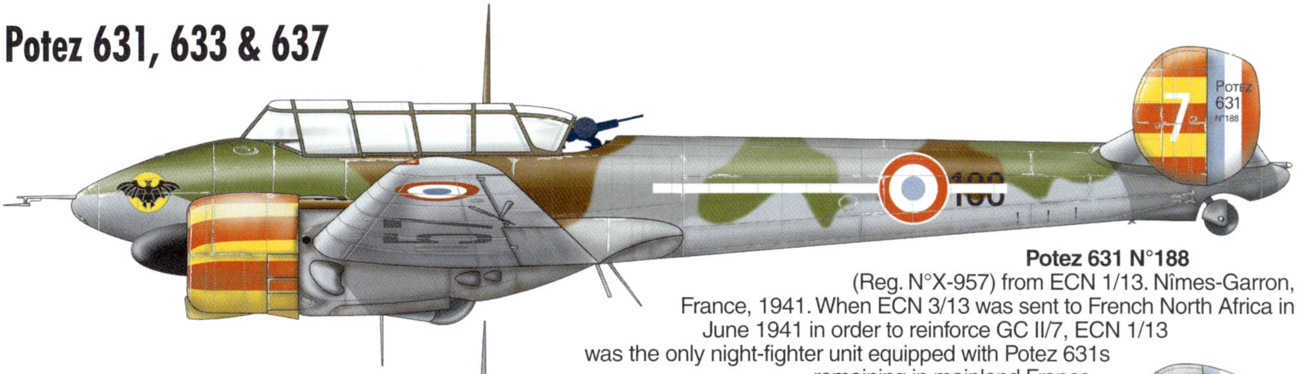

Potez 631 N°188
(Reg. N°X-957) from ECN 1/13. Nîmes-Garron, France, 1941. When ECN 3/13 was sent to French North Africa in June 1941 in order to reinforce GC II/7, ECN 1/13 was the only night-fighter unit equipped with Potez 631s remaining in mainland France.

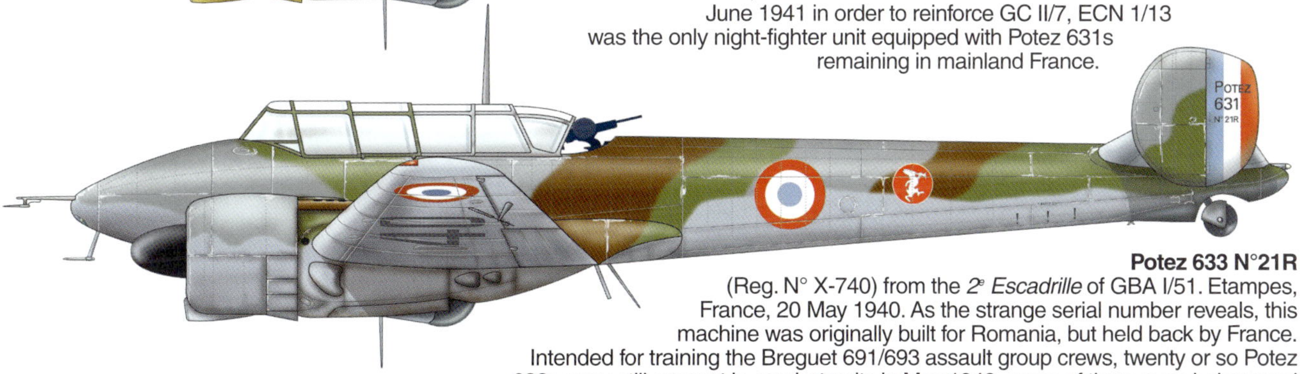

Potez 633 N°21R
(Reg. N° X-740) from the *2e Escadrille* of GBA I/51. Etampes, France, 20 May 1940. As the strange serial number reveals, this machine was originally built for Romania, but held back by France. Intended for training the Breguet 691/693 assault group crews, twenty or so Potez 633s were still present in combat units in May 1940, some of them even being used in operations at the beginning of the Battle of France.

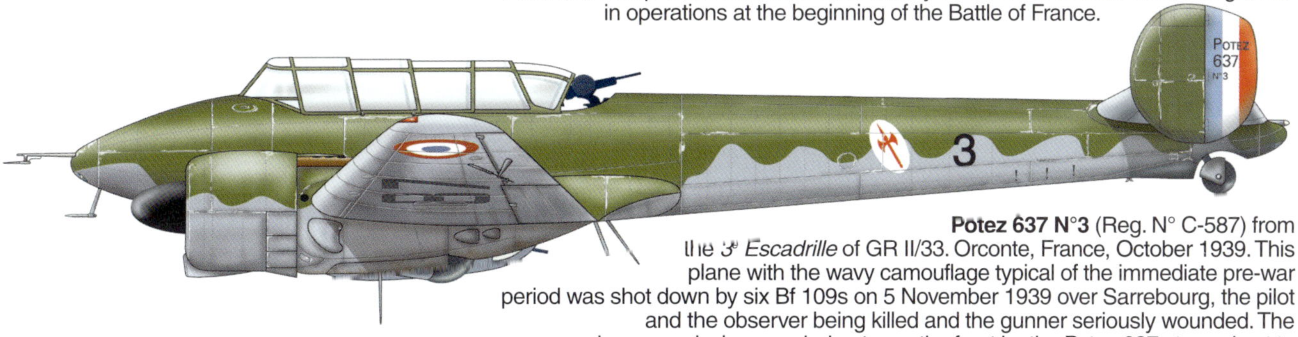

Potez 637 N°3 (Reg. N° C-587) from the *3e Escadrille* of GR II/33. Orconte, France, October 1939. This plane with the wavy camouflage typical of the immediate pre-war period was shot down by six Bf 109s on 5 November 1939 over Sarrebourg, the pilot and the observer being killed and the gunner seriously wounded. The reconnaissance missions carried out over the front by the Potez 637s turned out to be particularly deadly in the first weeks of fighting since the aircraft were particularly vulnerable without fighter cover.

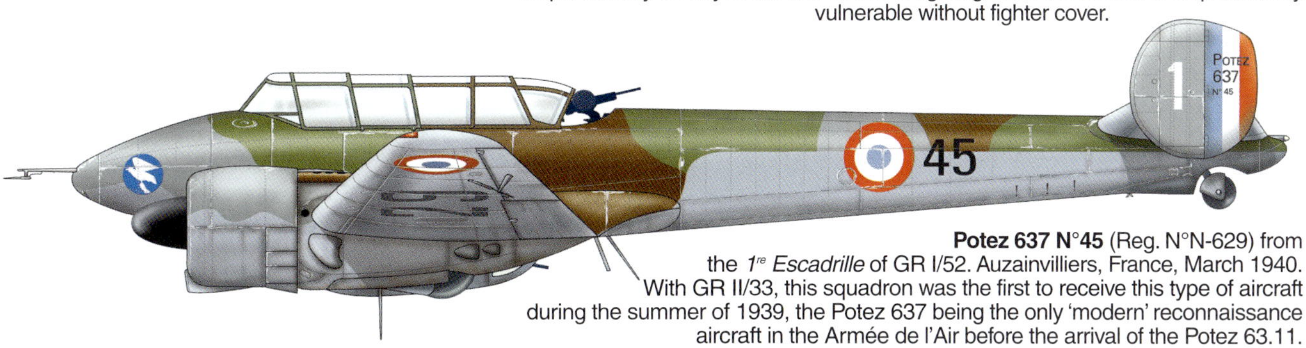

Potez 637 N°45 (Reg. N°N-629) from the *1re Escadrille* of GR I/52. Auzainvilliers, France, March 1940. With GR II/33, this squadron was the first to receive this type of aircraft during the summer of 1939, the Potez 637 being the only 'modern' reconnaissance aircraft in the Armée de l'Air before the arrival of the Potez 63.11. As a result it was the only one allowed to cross into enemy territory in day-time after the MB 131s were withdrawn.

Potez 63.11

The Potez 63.11 is no doubt the most emblematic plane of the dramatic Battle of France in May-June 1940. With more than 700 examples produced before the Armistice, it equipped indeed not only most of the Armée de l'Air's reconnaissance groups, but also a considerable number of GAOs, its crews having to make up for its defects with their courage and skill in order to carry out the missions they were given which if not all strategic were always very risky...

When the Potez 637 A3 was ordered by the Armée de l'Air during the summer of 1938, it was considered as a transition aircraft whilst waiting for a model with better performances intended to replace the Breguet 270s, Potez 39s and ANF-Les Mureaux to be found for the GAOs. The SNCAN therefore proposed a variant very closely derived from the twin-engined fighter version (keeping the wings, the engines, the undercarriage and the rear part of the fuselage), designated **Potez 63.11 A3** of which three prototypes were bought, the first taking to the air for the first time on 31 December 1938. The authorities' confidence in this prototype was such that 145 examples were ordered before the maiden flight.

The new machine was different in that it had a new redesigned front, with the observer now installed in a glazed nose and the pilot moved further back; the radio operator did not change places. As a result the canopy was shortened and access to the inside was by means of removable panels and trap doors. The fuel tanks were also spread out differently (two in the wing central section and two others in the outer sections). Armament comprised a machine gun installed under the belly firing forward and two others operated by the machine gunner, one flexible and the other firing through the floor and foot-operated, defending the upper and lower sectors respectively.

The first trials for N°01 having taken place with convex panes in the nose, the front of the plane was modified after the plane had been to the CEMA (flat panes to improve visibility), and lengthened.

As all the GAOs were to be re-equipped, 1 684 of the new machines were ordered for only 1939, production being divided out between the SNCAN factories situated at Meaux and Les Mureaux. Although the first production machine flew on 10 July 1939, production was nevertheless perturbed by difficulty supplying the Ratier propellers and some machines had to be delivered with wooden two-blade Levasseur propellers instead... Finally, the equipment depended on the type of engine used, the GR 14 M 4/5s taking the Ratier 1527/8 as planned, the 14 M 6/7 being fitted with Gnôme & Rhône 2190/1 propellers.

When the war started, with the Dewoitine D.720 'triplace de travail' (army co-operation three-seater) still on the drawing board, new orders for the Potez 63.11 were placed, 860 machines being put into service by May 1940 according to the plans (12 GAOs with 9 planes and 26 with 6) which were quantities requiring a sustained production rate of 150 planes per month.

The first unit to be equipped was the GR II/52 which received the new Potez 63.11s in November 1939 followed the same month by GR I/52 and II/33. In December it was the turn of GRs I and

Above:
Potez 63.11 from GR II/14. There are a lot of markings on this machine and they include the white Armistice stripe changed into an arrow, the serial number, the individual number (shaded) and finally the blazon containing the two insignia of the escadrilles making up the group. *(SHAA)*

Opposite:
For the tests, the engines of Potez 63.11 N° 91 have had their cowlings removed and the two-blade propellers have been installed without their spinners. Note that the anemometric tube which was normally in the nose on the 631s has been placed on the left-hand wing of the Potez 63.11. *(IWM)*

II/36 to receive their machines and the first operational sortie was on 13 February, made by the aircraft from GR I/33. During the winter of 1939-1940, conversion to the new machines carried on in the units, including also several GAOs (three machines being allotted to half a dozen of them in November 1939). A unit based at Toulouse was specially set up for this purpose in October 1939, followed by several training centres, so much so that on 10 May 1940, thirty or so GAOs were flying Potez 63.11s; this figure has to be put into perspective considering that some units were brought up to strength by using older machines like the Potez 637, different versions of the ANF-Les Mureaux, the Potez 25 or 390 and even the Breguet 270 and that half a dozen GAOs were in the process of being converted.

Below:
The observer was located in the glazed front and had a camera for taking vertical shots. The 7.5 machine gun under the fuselage was fired by the pilot and two others of the same calibre were sometimes added. In hot weather, the propeller spinners were removed to improve engine cooling. *(Wing Masters Archives)*

Above:
As with this machine seen after the Armistice, the armament of some Potez 63.11s was reinforced, namely two pods containing one 7.5 machine gun each installed under each wing and which could be complemented by additional weapons added under the fuselage, firing forward or rearwards, or even twin guns in the rear gunner's position. Armed like this the planes were intended for ground attack operations - an impossible task for an under-powered observation plane! *(Coll. J. Delmas)*

On 15 March 1940, 459 Potez 63.11s had been delivered and accepted by the Armée de l'Air, 217 of them fitted with GR 14 M 6/7 engines with GR propellers and 95 were powered by 14 M 4/5 engines; 147 had to be happy with two-blade wooden propellers. In all 294 of these machines found themselves in the front line, of which 109 in the reconnaissance units; only a small number had had their armament reinforced (four machine guns under the wings, three under the fuselage and twin guns in the rear cockpit).

A few weeks later when the Blitzkrieg started, about 70% of the GAOs were not operational due to lack of spare parts, a lot of them being destroyed on the ground during the first days of the campaign whereas certain units were withdrawn from the front line without even being sent into combat. The losses suffered by the Potez 63.11s during the Battle of France were heavy (the heaviest

TECHNICAL SPECIFICATIONS

Type
Three-seat twin-engined observation plane (A3)

Power plants
Two 14-cylinder Gnôme-Rhône 14 M 4/5 radial engines rated at 700 bhp each.

Dimensions
Length: *36 ft.*
Height: *10 ft 1 in.*
Wingspan: *52 ft 10 in.*
Wing Area: *351.950 sq ft.*
Unladen Weight: *6 897 lb.*
Max. Weight: *9 966 lb.*

Performance
Max. Speed: *296 mph at 18 150 ft.*
Range: *937 miles.*
Service Ceiling: *28 050 ft.*

Armament
Three 7.5-mm MAC 1934 machine guns under the fuselage, firing forward, one defending the upper sector, one the lower, foot-operated. (Some machines carried increased armament: two pods containing one 7.5-mm machine gun each under each wing; two extra machine guns under the fuselage; twin MAC machine guns in the rear position; two extra machine guns firing downwards.) Up to 440 lb of bombs on four underwing pylons.

Potez 63.11 N°225 photographed in 1940. Unlike the fighter versions (631), this type was not given the white identification stripe along the fuselage during the Battle of France.
(Coll. J. Delmas)

losses of all the planes in the Armée de l'Air) their missions being often very dangerous and carried out without fighter protection and against considerable enemy opposition (fighters and flak). Fortunately, the aircraft handled very well and was very robust which contributed to saving not a few crews... As an example of this dramatic atmosphere, let us mention the case of GAO 510 which lost twelve machines, of which six on the ground in six weeks after being stationed on fourteen different air fields.

The Armistice conventions demanded that a lot of units be disbanded: GR I/33, I/35, II/36, I/52 and I/55 as well as all the GAOs except for 1/583. Only GR I/14 at Perpignan, GR II/14 at Avignon and GR II/22 at Montpelier survived in mainland France, and GR I/22 (Rabat) and GR I/36 (Sétif) in French North Africa, together with GR II/39 based in Syria with GAO 1/583, these last two units being disbanded after taking part in the fighting in June-July 1941.

An extra squadron, GR I/52 recreated at Marrakech in September 1940 also received some Potez 63.11s, as did GR II/63 (which was at first equipped with Martin 167Fs, then with the Potez) after it set itself up at Bamako (Mali) in September 1941. Finally, some aircraft were sent to Madagascar in July 1941 where they were amalgamated with the MS 406s in the *Escadrille de Chasse 565* in a *Groupe d'Aviation Mixte* which lost four of its Potez in May 1942 when the island was attacked by Commonwealth forces.

In 1941, the occupying authorities allowed production of the machine to resume. 702 had been built by 31 May 1940, 431 built at Méaulte, 243 at Mureaux and 28 at Evreux, where the production line had been transferred because of the enemy's advance. The total number of Potez 63.11s built is not known (the highest serial number was 866) and must not be confused with the number of machines delivered to the Armée de l'Air which was less than the factory production rate.

Therefore about 120 machines were finished or built which the Luftwaffe used mainly for training its multi-engine aircraft crews; after the invasion of the 'Unoccupied Zone' some examples were handed over to Rumania in exchange for oil.

The Potez 63.11 started entering service when war was declared and hardly had the opportunity to give rise to any real variants which ended up with the building of a prototype, except for the Potez 63.16, a '*triplace de travail*' intended for training crews for army cooperation (T3) which was nothing but a Potez 63.11 with bigger wings. The only prototype flew in January 1940.

Below:
Rear defensive gunner's position on the Potez 63.11 had a flexible 7.5-mm machine gun and another operated by a foot pedal firing through the floor. The tube for the wire antenna aerial which was deployed in flight can be made out next to the stirrup step.
(WingMasters Archives)

Potez 63.11

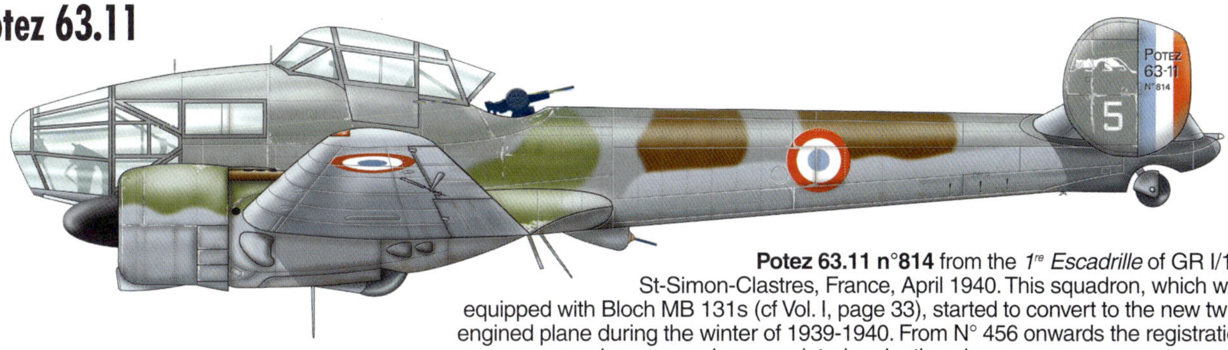

Potez 63.11 n°814 from the *1re Escadrille* of GR I/14. St-Simon-Clastres, France, April 1940. This squadron, which was equipped with Bloch MB 131s (cf Vol. I, page 33), started to convert to the new twin-engined plane during the winter of 1939-1940. From N° 456 onwards the registration number was no longer painted under the wings of the Potez 63.11s.

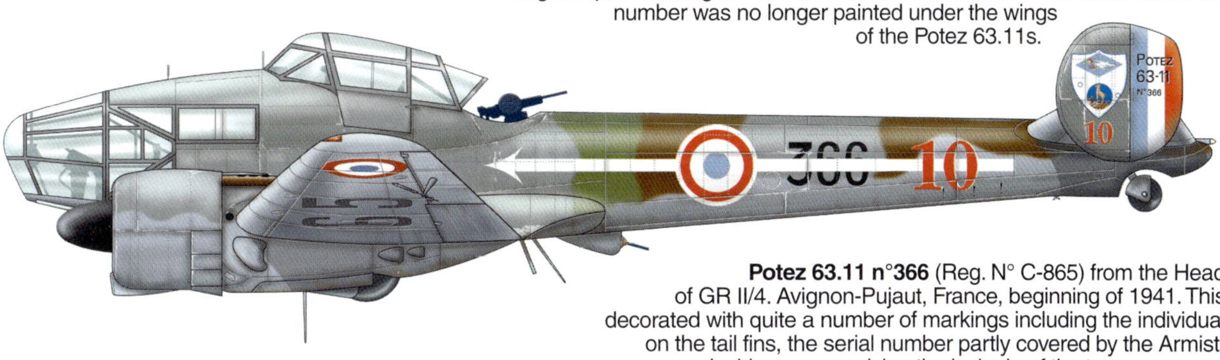

Potez 63.11 n°366 (Reg. N° C-865) from the Headquarters of GR II/4. Avignon-Pujaut, France, beginning of 1941. This plane is decorated with quite a number of markings including the individual number on the tail fins, the serial number partly covered by the Armistice stripe and a blazon comprising the insignia of the two escadrilles making up the group.

Potez 63.11 n°617 from the *2e Escadrille* of GR II/22. Montpelier-Fréjorgues, France, 1941. This plane, whose camouflage has been touched up and which bears the inscription '*Cne Massogi*' on the right of the pilot's canopy, is visible on this period photograph next to a machine bearing the yellow and red stripes of the Vichy air force, proof that the directives were not always applied in the same manner...

Potez 63.11 n°12 (Reg. N° J-815) from the *2e Escadrille* of GR II/22. Montpellier-Fréjorgues, France, beginning of 1942. This squadron was one of the three units equipped with this type of twin-engined aircraft in mainland France after the Armistice. In theory, the strength of each reconnaissance squadron was thirteen machines.

Potez 63.11

Potez 63.11 n°156 (Reg. N° C-655) from the 2e *Escadrille* of GR II/33. Athies-sous-Laon, France, April 1940. This machine was one of the first six 63.11s delivered to this unit.

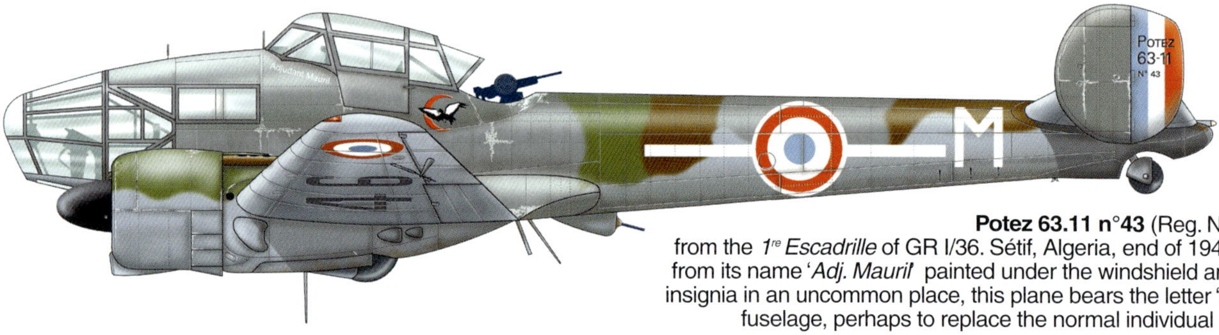

Potez 63.11 n°43 (Reg. N° J-846) from the 1re *Escadrille* of GR I/36. Sétif, Algeria, end of 1940. Apart from its name '*Adj. Mauril*' painted under the windshield and a unit insignia in an uncommon place, this plane bears the letter 'M' on its fuselage, perhaps to replace the normal individual number.

Potez 63.11 n°450 (Reg. N° C-949) from the 1re *Escadrille* of GR I/52. Marrakech, Morocco, April 1941. Initially, GR I/52 was disbanded after the Armistice, then re-formed in September 1940 and was one of the three GR units equipped with Potez 63.11s based overseas, together with I/36 at Sétif (Algeria) and II/63 at Thies (Senegal).

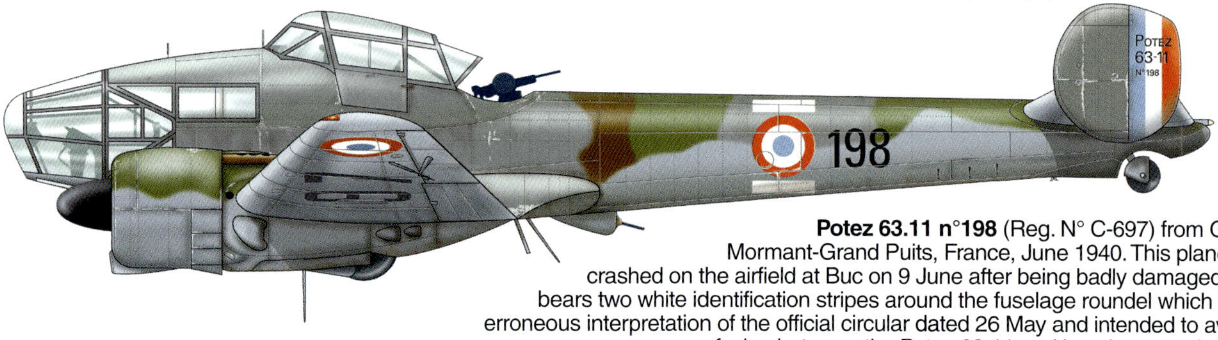

Potez 63.11 n°198 (Reg. N° C-697) from GR II/55. Mormant-Grand Puits, France, June 1940. This plane, which crashed on the airfield at Buc on 9 June after being badly damaged by flak, bears two white identification stripes around the fuselage roundel which were an erroneous interpretation of the official circular dated 26 May and intended to avoid any confusion between the Potez 63.11 and its adversary, the Bf 110.

Potez 63.11

Potez 63.11 n°641 from GAO 515. Connantre, France, June 1940. This plane was one of the rare machines to have borne a personal insignia, here 'Grumpy' from the cartoon *Snow White and the Seven Dwarves* which was very popular in France at the time, painted under the windshield. The GAO planes very rarely bore their unit insignia using their serial number more often, painted white on the fuselage as an individual number.

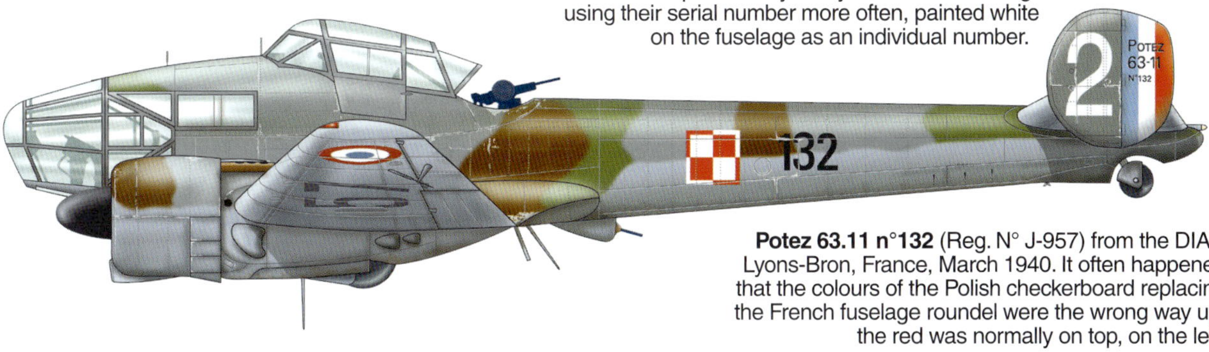

Potez 63.11 n°132 (Reg. N° J-957) from the DIAP. Lyons-Bron, France, March 1940. It often happened that the colours of the Polish checkerboard replacing the French fuselage roundel were the wrong way up: the red was normally on top, on the left.

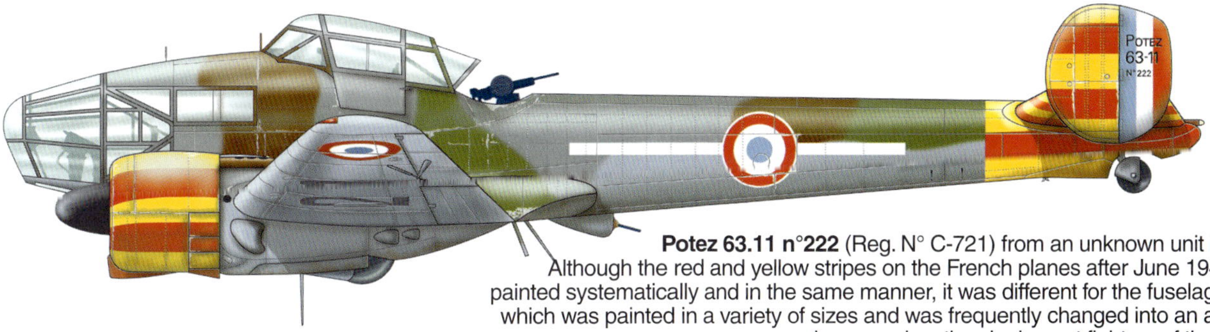

Potez 63.11 n°222 (Reg. N° C-721) from an unknown unit in 1941. Although the red and yellow stripes on the French planes after June 1941 were painted systematically and in the same manner, it was different for the fuselage stripe which was painted in a variety of sizes and was frequently changed into an arrow as happened on the single-seat fighters of the period.

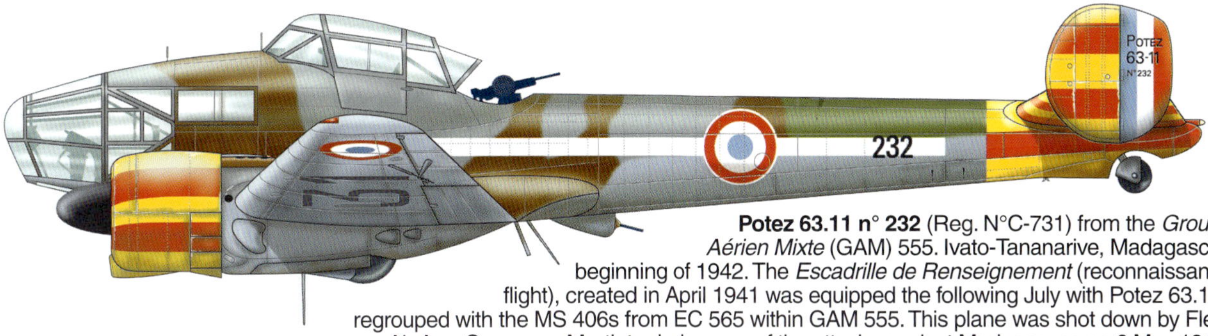

Potez 63.11 n° 232 (Reg. N°C-731) from the *Groupe Aérien Mixte* (GAM) 555. Ivato-Tananarive, Madagascar, beginning of 1942. The *Escadrille de Renseignement* (reconnaissance flight), created in April 1941 was equipped the following July with Potez 63.11s regrouped with the MS 406s from EC 565 within GAM 555. This plane was shot down by Fleet Air Arm Grumman Martlets during one of the attacks against Madagascar, on 6 May 1943.

THOSE THAT NEVER WERE...

The formula invented by the engineer Nenadovitch during the twenties whereby the second wing of a biplane was moved aft in order to get it out of the slipstream of the first no matter what the aircraft's behaviour, was taken up by Maurice Delanne. He adapted it by using the rear wing as the horizontal tail surfaces with, in this case, the addition of two fins. The front wing was higher and gull-wing shaped. This strange set up was tried out on civilian planes and was supposed to enable the plane to be more manoeuvrable and have a very low stalling speed, and to give the rear gunner an exception field of vision. French GHQ expressed doubts about the feasibility of such a formula, but a prototype was nevertheless decided upon and manufacturing entrusted to the Arsenal de l'Aéronautique, a nationalised company, whose statutory obligations included help for the 'little' aircraft builders. It was the scarcely convincing C2 category ('*Biplace de Combat*' - Combat two-seater) which was chosen as the specification and it gave rise to the **Arsenal-Delanne 10**. This was a machine which combined modern techniques (metal stressed-skin construction, camber flaps, leading edge slats, undercarriage retracting into the fuselage) and an archaic design (biplane). Moreover, the cockpit was placed right at the rear, meaning poor forward visibility for the pilot but giving an excellent field of vision for the machine gunner sitting at the very rear of the fuselage, in a turret. The prototype's first flight was delayed (the tail wheel was only available on the 10 May 1940) and the machine was finally ready just when the Germans arrived. The first flight did not take place until October 1941; the armament was never installed. The Delanne 10 was sent to Germany in 1943, but it is not known if any tests were carried out there; if they were they must have been disappointing as the lay-out chosen was rather hybrid and not really workable.

Arsenal-Delanne 10

TECHNICAL SPECIFICATIONS

Type
Two-seat fighter (C2)
Power Plant
One 12-cylinder Hispano-Suiza 12 Ycrs in-line liquid-cooled engine rated at 860 bhp at take-off.
Dimensions
***Length:** 24 ft.*
***Height:** 9 ft 11 in.*
***Wingspan:** 34 ft 7 in.*
***Wing Area:** 242.167 sq ft.*
***Loaded Weight:** 6 270 lb.*

Performance
***Max. Speed:** 344 mph.*
***Endurance:** 1h.30.*
***Service Ceiling:** 33 000 ft.*
Armament (Proposed):
One engine-mounted 20-mm Hispano-Suiza cannon, two wing-mounted 7.5-mm MAC 34 machine guns; rear armament unknown.

Dewoitine D.720

For the T3 programme (three-seat working aircraft) issued in 1937 concerning a twin-engined multi-role aircraft capable of carrying out reconnaissance, army cooperation, and light bombing missions, the SNCAM into which the Dewoitine company had been incorporated when the armament industry was nationalised, proposed its **D.700** in January 1938.

This was a monoplane with fixed undercarriage and a high cantilever wing under which the engine nacelles were attached.

Construction of the prototype had just started when the plane was redesigned, its fuselage changed (glazed-in nose), new engines fitted, and retractable undercarriage installed.

These structural modifications meant a re-designation of the machine which then became the **D.720.** The only prototype flew for the first time on 10 July 1939 and was sent to the CEMA the following September for official tests. Although the Potez 63.11 had started to enter service and a specialised variant, the Potez 63.16 T3, had even started its trials, some thousand examples of the D.720 were ordered, to be delivered at the end of 1940.

Finally the dramatic turn events had the best of the project, the prototype disappearing definitively in the upheaval following the Armistice in June 1940...

TECHNICAL SPECIFICATIONS

Type:
Reconnaissance three-seater (T3)
Power plants:
Two Renault 12 R 00/01 air-cooled V-12 engines rated at 500 bhp each.
Dimensions
***Length:** 34 ft 7 in.*
***Height:** 12 ft 4 in.*
***Wingspan:** 49 ft 1 1/2 in.*
***Wing Area:** 295.982 sq ft.*
***Loaded Weight:** 8 393 lb.*

Performance
***Max. Speed:** 225 mph at 11 880 ft.*
***Range:** 956 miles*
***Service Ceiling:** 27 720 feet.*
Armament
Four MAC 7.5-mm machine guns (two in the turret, one under the fuselage firing forward and one firing downwards, through the fuselage). 440 lb of bombs.

Dewoitine D.770

Following on the AB2 (assault two-seat aircraft) programme issued in August 1937 which gave rise to the Breguet 691 (cf. Vol. 1 p.58) the SNCAM proposed their **D.770**, a twin-engined aircraft equipped with liquid-cooled Hispano-Suiza engines and the **D.771** which was identical but powered by air-cooled Gnôme-Rhône engines.

It was a cantilever monoplane, with a monocoque fuselage to which only the nose was added: it contained the offensive weapons. The crew was well protected by armour; the bomb bay could carry eight 110-lb bombs. A prototype of each version was ordered officially, the D.770 N°01 flying for the first time at the end of July 1939. The D.771, although it was completed, never actually flew, since the MB 175 was chosen in preference to it and the authorities decided to reduce the number of different types that were being produced. The D.770 prototype finished its initial trials but never had the time to be sent to the CEMA because of the Armistice. The two prototypes were finally scrapped in 1941.

TECHNICAL SPECIFICATIONS

Type
Assault three-seater (AB3)
Power plants
Two 12-cylinder water-cooled Hispano-Suiza 12 Yjrs/krs engines rated at 1 050 bhp on take-off.
Dimensions
Length: *32 ft.*
Height: *10 ft 6 in.*
Wingspan: *49 ft 10 in.*
Wing Area: *317.508 sq ft.*
Loaded Weight: *13 409 lb.*
Performance
Max. Speed: *312 mph at 10 560 ft.*
Service Ceiling: *27 225 ft.*
Armament
Two HS 404 20-mm cannon and one 7.5-mm MAC machine gun firing forward; one 7.5-mm machine gun in the rear upper gun position and two in the lower; 880 lb of bombs.

NC 150

In 1938, the SNCAC, created by the merger of the Farman and Hanriot firms presented two projects with a fully-pressurised cockpit. The first, the biggest, was the **NC 140** which used the wings, the tail and the undercarriage of the NC 223.3/4, but this was never built. Production of the second, designated **NC 150,** started in 1938 with pressurisation tests being carried out on another of the firm's planes, the **NC 130**. While waiting for the results of the tests, it was decided to build two non-pressurised prototypes of the bombers, this feature only being added on the third prototype. The construction design used as many non-strategic materials as possible (especially wood). The general concept was given the nick-name 'bi-tri' because a third engine, installed in the fuselage and cooled by means of scoops fitted under the belly, was only used to drive the engine superchargers positioned as usual in the wings.

The first prototype, NC 150.01, took off for the first time on 11 May 1939, powered by two Hispano-Suizas boosted by an NC-C1 supercharger driven by an engine placed over the bomb bay. A good number of modifications and improvements (glazed nose, fins and turbochargers) were made during the trials.

Because of the worsening military situation and while production of the LeO 451s and Amiot 351/354s was suffering from severe shortages of light alloys, the SNCAC was given the order to abandon the pressurised 'bi-tri' concept in favour of a simple twin-engined version and to build a series of machines equipped either with supercharged Hispano-Suiza 12 Y engines, designated **NC 152**, or with Pratt and Whitney Double Wasps, the **NC 153**.

The definitive version of the bomber, represented by the third prototype, the NC 151.03, was in fact never finished, the NC 150.01 being evacuated to Bordeaux a short while before the end of hostilities where it was more likely than not destroyed.

TECHNICAL SPECIFICATIONS

Type
Four-seat bomber (B4)
Power plants
Two Hispano-Suiza 12 Y 32-33 engines rated at 955 bhp each at 29 700 ft boosted by a supercharger driven by an HS 12 Xirs engine rated at 690 bhp.
Dimensions
Length: *58 ft*
Wingspan: *72 ft 1 in.*
Wing Area: *661.924 sq ft.*
Loaded Weight:
22 169 lb.
Performance
Max. Speed:
675 mph at 26 400 ft.
Range: *1 375 miles*
Service Ceiling:
37 950 feet.
Armament
One 20-mm HS 404 canon in the dorsal gunner's position, one 7.5-mm in the ventral gondola and another in the nose. 2 970 lb of bombs.

Hanriot NC 530

For the 1936 T3 programme, a team of engineers from the Hanriot firm at the time incorporated into the SNCAC proposed the **NC 150**, a high wing monoplane with fixed undercarriage whose only prototype took to the air for the first time in June 1938.

One of the inconveniences of the machine was its ventral gondola which accommodated the observer in a position which was dangerously exposed to enemy fire and at the same time greatly increased drag. The original design was therefore turned down and gave rise to the NC 530 Exp (for experimental) designed as a test bench. It was only at the end of 1939 that the final prototype, the NC 530 N°01, started its trials. The new machine was built entirely of metal, had a different wing from the two previous prototypes, a larger glazed nose and a bomb bay which could house bombs or cameras, or even an extra fuel tank. The trials carried on into May 1940, but vibration problems in the tail which had appeared at the outset were never entirely ironed out. Moreover, at the time, the Potez 63.11 had become the standard twin-engined reconnaissance aircraft in the Armée de l'Air which precluded any production series being built. The NC 530 prototype ended its career without glory, destroyed in 1940 as it was being evacuated towards the South of France.

Technical Specifications

Type
Reconnaissance three-seater
Power plants
Two fourteen-cylinder Gnôme-Rhône 14 m radial engine rated at 700 bhp each.
Dimensions
Length: *37 ft 5 1/2 in.*
Wingspan: *43 ft 1 in.*
Wing Area: *68 sq 7 ft.*
Loaded Weight: *11 209 lb.*
Performance
Max. Speed: *256 mph at 16 995 ft.*
Range: *1 125 miles*
Service Ceiling: *28 050 feet.*
Armament
Two 7.5-mm machine guns, 1 100 lb of bombs.

Hanriot NC 600

In order to respond to an official request in 1934 for a twin-engined long-range escort fighter, bomber interceptor and ground attack aircraft, Hanriot presented its **H 220** at the 1938 Salon de l'Aviation. Modified before its first flight, fitted with new engines and a new twin tail, the prototype was re-designated Hanriot **H 220-1** and finally flew for the first time on 2 September 1937. Having been destroyed in an accident the following February, work was started on a new prototype, first christened **Hanriot H 220-2** then **NC 600** N°01 because the firm was nationalised and incorporated into the SNCAC followed by an official order for a further five models. The new prototype was not different from the previous one except for a modified wing. It kept the entirely metal construction, the mid-wing reinforced by two wing pylons and the ovoid airframe comprising two individual compartments separated by a fuel tank. The armament consisted of two 20-mm cannon and three machine guns, one of which partly retracted into the belly of the machine. Having flown for the first time on 15 May 1940, the NC 600 N°01 very quickly revealed its qualities, its low-level handling being the only thing that needed improving. A second prototype (NC 600 N°02) was ready to take to the air in its turn when the SE 100 was chosen in its place and it was decided to use the Bourges factory for Breguet 696 production (cf.Vol. I p. 60.). The German occupation which started a few days later put paid to all these projects.

Technical Specifications

Type
Two-seat fighter
Power plants
Two fourteen-cylinder Gnôme-Rhône 14 M 00/01 radial engines rated at 700 bhp.
Dimensions
Length: *29 ft.*
Wingspan: *42 ft 3 in.*
Wing Area: *227.637 sq ft.*
Loaded Weight: *8 789 lb.*
Performance
Max. Speed: *338 mph at 16 500 ft.*
Range: *537 miles*
Max. Ceiling: *26 400 feet.*
Armament
Two 20-mm HS 404 cannon firing forward; two MAC 7.5-mm machine guns and one 20-mm canon in the rear gunner's position.

Potez 220

In the programme which gave rise to the Bloch MB 174/175 and to which the Glenn Martin 167F was added, the Les Mureaux company, nationalised in 1937 as well as the Potez factories and incorporated into the SNCAN, presented the **Potez 220** (the name of the aircraft builder was retained whatever the project because it had considerable prestige at the time, particularly internationally) a 'Triplace de travail' (Army co-operation three-seater) whose design was less elaborate than its competitor, the Bloch, since the observer was accommodated in a ventral gondola which reduced the aircraft's performances and seriously exposed the occupant to enemy fire. The plane was built entirely of metal with the fuel tanks all located in the wings. Two prototypes were built, distinguished mainly by their power plants. The Potez 220 N°01 which flew for the first time on 7 April 1939 was in fact